AF508679

Advances and Trends in Non-conventional, Abrasive and Precision Machining

Advances and Trends in Non-conventional, Abrasive and Precision Machining

Editors

Mariusz Deja
Angelos P. Markopoulos

MDPI • Basel • Beijing • Wuhan • Barcelona • Belgrade • Manchester • Tokyo • Cluj • Tianjin

Editors

Mariusz Deja
Gdańsk University of
Technology
Poland

Angelos P. Markopoulos
National Technical University of
Athens
Greece

Editorial Office
MDPI
St. Alban-Anlage 66
4052 Basel, Switzerland

This is a reprint of articles from the Special Issue published online in the open access journal *Machines* (ISSN 2075-1702) (available at: https://www.mdpi.com/journal/machines/special_issues/Abrasive_Precision_Machining).

For citation purposes, cite each article independently as indicated on the article page online and as indicated below:

LastName, A.A.; LastName, B.B.; LastName, C.C. Article Title. *Journal Name* **Year**, *Volume Number*, Page Range.

ISBN 978-3-0365-0602-9 (Hbk)
ISBN 978-3-0365-0603-6 (PDF)

Contents

About the Editors

Mariusz Deja is an Associate Professor and the Head of the Department of Manufacturing and Production Engineering at the Faculty of Mechanical Engineering and Ship Technology, Gdańsk University of Technology, Poland. His research interests are abrasive finishing processes, advanced CNC programming, computer-aided process planning using feature-based modeling, as well as additive technologies. He is an author of more than 100 scientific papers in journals, conference proceedings, and book chapters showing the current research in the area of advanced manufacturing processes.

Angelos P. Markopoulos is an Assistant Professor in the Laboratory of Manufacturing Technology at the School of Mechanical Engineering, National Technical University of Athens, Greece. His research includes topics such as precision and ultraprecision conventional and nonconventional machining processes, with a special interest in advanced manufacturing and Industry 4.0. Furthermore, he is an expert in manufacturing technology modeling and simulation, including the finite element method, artificial intelligence, and molecular dynamics. He is the author of more than 120 papers in journals, conferences, and book chapters on the abovementioned areas and a member of the editorial boards of several international journals.

Preface to "Advances and Trends in Non-conventional, Abrasive and Precision Machining"

The modern, highly competitive industrial environment demands machining and production processes that result in exceptional quality and precision. The general trend to design and manufacture more complicated mechanical components, along with the rapidly forward-moving material science, raises the need to incorporate and develop new machining techniques in the manufacturing process. Nonconventional machining processes differ from conventional ones, as they utilize alternative types of energy, such as thermal, electrical, and chemical energy, to form or to remove material. Commonly, the energy source has high power density, while the process features prodigious accuracy and the capability to produce and handle demanding shapes and geometries. Examples of nonconventional machining processes are electrical discharge machining (EDM), electrochemical machining (ECM), laser processing, and laser-assisted machining. Abrasive processes such as grinding, lapping, honing, polishing, and superfinishing are constantly developing and allow for obtaining a fine surface finish along with high efficiency. There is an increased scientific and commercial interest in the in-depth understanding and further development of the aforementioned nonconventional and precision machining processes. Research is moving forward through experimental studies, as well in the field of modeling and simulation, exploiting the increased available computational power. Multiphysics and multidisciplinary and multiscale modeling are powerful tools in the effort to optimize existing nonconventional precision machining processes, as well to develop novel ones. As their wider use by the industry swiftly grows, research has to be focused on them, not only due to the academic and scientific interest, but also for the possible financial gain. In light of these aspects, this book contains recent advances and technologies in the aforementioned fields, indicating the future trends for nonconventional precision machining processes. More specifically, a work where brushing with bonded abrasives as the finishing process for ceramics is presented [1]. The processing of zirconium dioxide workpieces with brushing tools of polycrystalline diamond-bonded grains is considered. The goal of the investigation is the reduction of grinding-related surface defects, the preservation of surface roughness and workpiece form, and the evaluation of tool wear in the case of brushing ceramic materials. It was found, by microscopical and surface topography measurements, that the brushing velocity and the grain size play the most significant role. Considering that the material removal mechanisms of abrasive brushing ceramics are largely unknown, this work is one of the few to deal with the specific topic. In the second work [2], a honing cell incorporating a thermographic camera, a sound intensity meter, and software for collecting and analyzing data received during the process on a CNC honing machine is proposed. With the aforementioned arrangement, images from the thermographic camera are analyzed online and the level of sound intensity obtained during honing is continuously monitored, with the purpose of online control of the process and its optimization. For further reducing the temperature of the workpiece due to its interaction with the tool and the subsequent deformations, the machining cell may have an automatic selection of the grain trajectory shape, with a specified value of the curvature radii of the abrasive grain trajectories, according to the wall thickness of the honed workpiece. With the proposed scheme, it is possible to increase the efficiency of the process by about 20-fold. The next work [3] also pertains to honing and more specifically to the possibility of employing wavelet analysis in order to evaluate the changes in the geometrical structure of a surface arising when honing with whetstones of variable granularity. The basics of the wavelet analysis and the differences between filtering with standardized filters, Fourier analysis, and the analysis of the results obtained when measuring the surface roughness with other wavelets are described. As a case study, the honing of a four-cylinder combustion engine was used, where roughness measurements of 3D spatial structures of the machined liners were carried out. The result of the work is the creation of basic recommendations for the selection of wavelets when assessing honed surfaces with different degrees of regularity of the traces that were generated on them. The effect of burnishing on heavily loaded structural elements operating in a corrosive environment is the subject of [4]. This research presents a fatigue resistance test performed on elements operating in seawater, namely a ship propeller, where various processing parameters of burnishing applied on samples are compared to specimens with a ground surface. The results indicate a 30% higher fatigue strength in a seawater environment of the burnished specimens. A device that allows for simultaneous turning and shaft burnishing with high slenderness is also presented. This device can be connected to the computerized numerical control system and automatic processes executed according to the machining program, with the aim of reducing the number of operations and cost of the process. Single-sided lapping is considered in [5], as it is one of the most effective planarization technologies. The process has relatively complex kinematics, and prediction of the tool wear is critical for product quality control. To determine the profile wear of the lapping plate, a computer model that simulates abrasive grains' trajectories in MATLAB is included in this work. Additionally, a data-driven technique was investigated to indicate the relationship between the tool wear uniformity and lapping parameters. The next two papers pertain to wire electrical discharge machining (WEDM) and die-sinking electrical discharge machining (EDM). The former [6] is a precise and efficient nonconventional manufacturing solution in various industrial applications, mostly involving the use of hard-to-machine materials such as, among others, the Inconel super alloys. The related study focuses on exploring the effect of selected control parameters, including pulse duration, pulse-off time, and the dielectric flow pressure on the WEDM process performance characteristics of Inconel 617 material. Parameters such as the volumetric material removal rate, dimensional accuracy of cutting, and surface roughness are considered in an experimental work that was carried out with the Box–Behnken design scheme and analyzed through the response surface methodology analysis of variance (ANOVA) tests. The latter study [7] presents an experimental investigation of the EDM of the aluminum alloy Al5052. A full-scale experimental work was carried out, with the pulse current and pulse-on time being the varying machining parameters. Then, polishing and etching of the perpendicular plane of the machined surfaces was performed in order to observe and measure the machined surfaces. Through analysis of variance (ANOVA), conclusions were drawn about the influence of machining conditions on the EDM performance, with consideration of the material removal rate, the surface roughness, the average white layer thickness, and the heat-affected zone microhardness. In the last paper [8], an analysis of the surface texture of turned parts with length/diameter ratios of 6 and 12 and various rigidity values is presented. The study pertains to samples of S355JR steel and AISI 304 stainless steel, with a detailed analysis of 2D surface profiles, using a large number of parameters that allowed significant differences in the surface microgeometry to be distinguished. The obtained results indicated significantly better roughness and waviness values of the AISI 304 steel surface in terms of its size, periodicity, and regularity; the turning process of AISI 304 shafts with low rigidity allows a better-quality texture to be achieved and has a positive effect on the general properties of a workpiece. Furthermore, it was concluded that the shafts with an L/D ratio of 12 had worse surfaces in the first two sections due to lower rigidity, while the results close to the three-jaw chuck, regardless of the L/D ratio and material type, demonstrated similar waviness and roughness parameters and profiles. The guest editors of this Special Issue would like to thank the authors for their valuable and high-quality work submitted, the reviewers for their efforts and time spent in order to improve the submissions, and the publisher for their excellent work and cooperation.

Mariusz Deja, Angelos P. Markopoulos
Editors

Article

Surface Finishing of Zirconium Dioxide with Abrasive Brushing Tools

Eckart Uhlmann [1,2] **and Anton Hoyer** [1,*]

[1] Institute for Machine Tools and Factory Management, Technical University Berlin, Pascalstr. 8-9, 10587 Berlin, Germany; eckart.uhlmann@iwf.tu-berlin.de

[2] Fraunhofer Institute for Production Systems and Design Technology, Technical University Berlin, Pascalstr. 8-9, 10587 Berlin, Germany

* Correspondence: hoyer@iwf.tu-berlin.de; Tel.: +49-30-314-22718

Received: 18 November 2020; Accepted: 13 December 2020; Published: 21 December 2020

Abstract: Brushing with bonded abrasives is a finishing process which can be used for the surface improvement of various materials. Since the machining mechanisms of abrasive brushing processes are still largely unknown and little predating research was done on brushing ceramic workpieces, within the scope of this work technological investigations were carried out on planar workpieces of MgO-PSZ (zirconium dioxide, ZrO_2) using brushing tools with bonded grains of polycrystalline diamond. The primary goal was the reduction of grinding-related surface defects under the preservation of surface roughness valleys and workpiece form. Based on microscopy and topography measurements, the grain size s_g and the brushing velocity v_b were found to have a considerable influence on the processing result. Furthermore, excessive tool wear was observed while brushing ceramics.

Keywords: abrasive brushing; finishing; fine machining; grinding; ceramics; MgO-PSZ; ZrO_2

---

1. Introduction

In order to meet increasing resource and productivity demands, modern technology requires the development of sustainable and responsible manufacturing processes. This leads to higher expectations in terms of component performance as well as durability [1]. Both are especially impacted by surface friction, which decreases the performance through energy loss and the durability through material wear. Friction between surface pairings is mainly influenced by their respective surface roughness, with high roughness leading to high friction, heat generation, and wear. However, the absence of surface roughness may also lead to a loss of retaining volumes for lubricant fluids. Therefore, surface finishing technologies no longer target total roughness reduction, but instead the ability to manufacture specialized surface textures for given tasks, even partially maintaining topography features. This enables, for example, the utilization of existing roughness valleys for lubricant retainment while only removing roughness peaks to further decrease surface friction.

Exemplary applications which require low surface roughness and high durability under frequent and selectively large pressures are artificial dentures or hip joints made of ceramic materials such as zirconium dioxide (ZrO_2) [2,3]. Ceramics are distinguished by high hardness, heat and wear resistance, and bio-compatibility [4]. Nonetheless, their generally low heat conductivity, high brittleness, and predisposition towards fracture formation make the machining of ceramics particularly challenging [5]. After the sintering, ceramics are typically brought into final shape by microcutting or microgrinding processes, both of which may lead to local surface defects and high surface roughness. Despite the relatively ductile machining characteristics of ZrO_2 compared to those of other types of ceramics, Fook and Riemer observed "brittle intercrystalline breakouts" as well as ductile deformation bulges while describing ground ZrO_2 surfaces [6]. These features likely affect the surface roughness adversely.

One machining process, which may be used for the finishing of components with hard surfaces, is brushing with bonded abrasives, making use of circular abrasive brushing tools (Figure 1). During tool production, abrasive grains are bonded in a filamentary polymer matrix by an extrusion process [7]. Common grain types are silicon carbide (SiC) and the softer aluminum oxide (Al_2O_3), although the finishing of ceramic materials suggests the use of harder grains such as polycrystalline diamond (PCD). As a polymer matrix, polyamide (PA) is most popular due to its high abrasion resistance, strength, as well as chemical inertness. Its thermal stability can further be increased by augmentation with additives [8]. The extruded abrasive filaments are cut to length and then attached to a circular brush body by either bonding or stuffing. However, largest filament quantities can be achieved by casting an epoxy resin brush body into a mold after arranging the filaments in a circular fashion. The thereby achieved high denseness of abrasive filaments allows for a high tool stiffness, increasing the productivity [7].

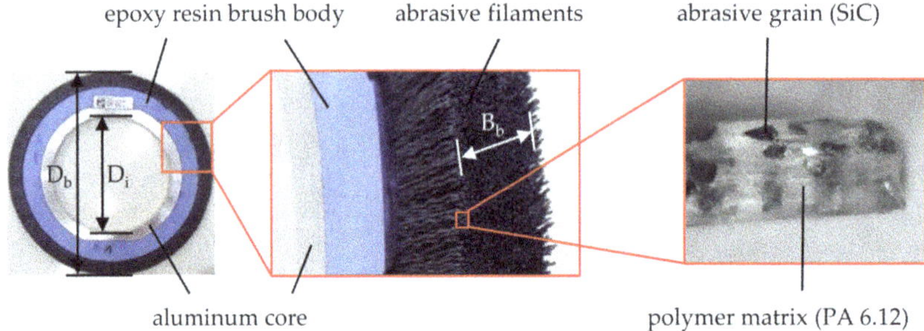

Figure 1. Layout of a circular abrasive brush tool [9].

Among the potentials of brushing with bonded abrasives are low process forces and temperatures as well as the utilization of existing machine systems, such as industrial robots, grinding machines, or mills [1,7]. Furthermore, the high flexibility of the abrasive filaments causes them to deflect during contact with a workpiece, thus assuming its shape and compensating minor inaccuracies regarding the geometry of workpiece, tool, and path [10]. Tool wear leads to a slow but successive shortening of the abrasive filaments [8,11,12], until the filaments become too short and therefore too stiff in order to yield consistent processing results. In addition, brushing tools typically require an initial conditioning process to achieve approximately equal filament lengths, which may be time- and cost-intensive, depending on the characteristics of the brushing tool and the requirements of the brushing process.

While abrasive brushing is widely used in industries, process designs are generally based on empirical values. This is mainly attributed to insufficient knowledge of the complex movement behavior of the flexible filaments, complicating the prediction of process characteristics, processing results, and tool wear. Thus, the technological investigations discussed in this article are directed at the gain of knowledge of brushing ceramics with bonded abrasives, specifically the characteristics of brushed surfaces in regard to relevant tool specification and process parameters.

2. Materials and Methods

To gain a basic understanding of the mechanisms being effective while processing ceramic surfaces with abrasive brushing tools, experiments were conducted on rectangular ZrO_2 workpieces of the type Frialit FZM, manufactured and ground to dimensions of $200 \times 200 \times 20$ mm^3 by FRIATEC AG, Mannheim, Germany. The sintered ZrO_2 matrix was partially stabilized (PSZ) with magnesium oxide (MgO) to strengthen the material by retaining cubic fluorite crystal structures—usually only present at temperatures T of >2300 °C—and thus prevented monoclinic crystal formation, which would lessen the fracture toughness [4]. Through a conventional plane grinding process, the surface roughness was then reduced to an arithmetic mean deviation of the roughness profile of Ra = 1.1 μm (Figure 2), with all roughness measurements taken orthogonally to the grinding direction.

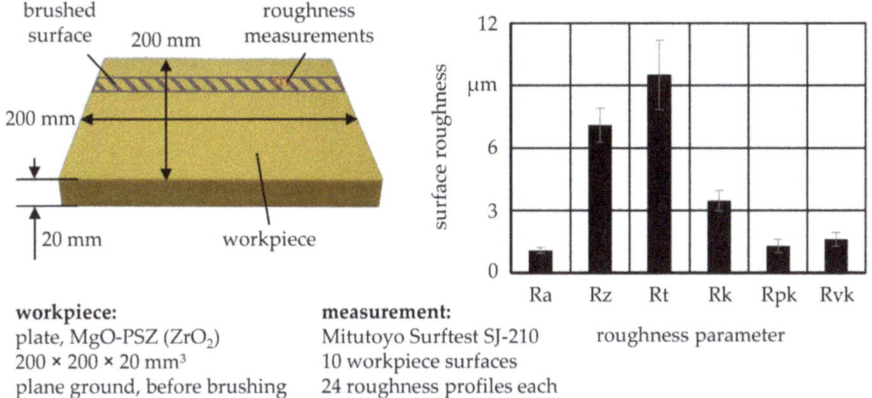

Figure 2. Workpiece properties.

The circular brushing tools were manufactured by C. Hilzinger-Thum Gmbh, Sindelfingen, Germany (Figure 3d). The tools featured an epoxy resin brush body with an outer diameter of D_b = 360 mm, an inner diameter of D_i = 203.2 mm, and a width of B_b = 20 mm, holding abrasive filaments with PCD grains bonded in a PA 6.12 matrix. Varied specification parameters include the grain size s_g and the filament diameter d_f. While the filament length l_f also impacts the processing results, too short filaments are prone to take thermal damage due to the high stiffness and the low heat conductivity of ceramic workpieces, as the filament plastic matrix has a relatively low melting temperature of $T_m \approx 150$ °C [8,13] and requires heat losses to be absorbed almost entirely by the workpiece. The accidental melting of the abrasive filaments can be compensated by the use of cooling lubricant [13]. However, PA 6.12 has a high tendency to absorb liquids, which decreases the filament stiffness and influences the consistency of processing results [9]. Therefore, no cooling lubricant was used during the technological investigations and a filament length l_f = 40 mm was chosen for all experiments, although further research needs to be done on both cooling lubricants and appropriate filament lengths l_f for the surface finishing of ZrO₂.

The technological investigations were carried out on a plane and profile grinding machine of the type Profimat MT 408 HTS, manufactured by Blohm Jung GmbH, Hamburg, Germany (Figure 3a). It was equipped with three linear axes, the encoders of which allowed for a positioning accuracy of approximately 1 μm. The spindle had a maximum performance of P_{max} = 45 kW and a maximum rotational speed of n_{max} = 11,000 min⁻¹. All workpieces were brushed in parallel to the direction of the preceding grinding process.

Before conducting the technological investigations, the machine tool was used to condition the brushing tools by shortening the filaments to approximately equal lengths l_f. As shearing off protruding filament tips with a sharp edge is a fast, yet imprecise method, a grindstone was used instead. After the radial runout error ΔD_b of the brushing tool fell below 0.4 mm, 0.11% of the brushing tool diameter D_b, an initialization process was repeatedly run with a brushing velocity of v_b = 20 m/s, a feed velocity of v_{ft} = 200 mm/min, and a penetration depth of a_e = 1 mm, hereinafter referred to as standard process parameters, until the reduction of the surface roughness did not significantly change any more, indicating quasi-static material removal behavior, which was required for consistent and comparable processing results. Subsequent experiments were conducted to determine the influence of the process parameters brushing velocity v_b, feed velocity v_{ft}, and penetration depth a_e on the workpiece surface roughness (Table 1), as well as the workpiece form deviation, especially after successive brushing cycles.

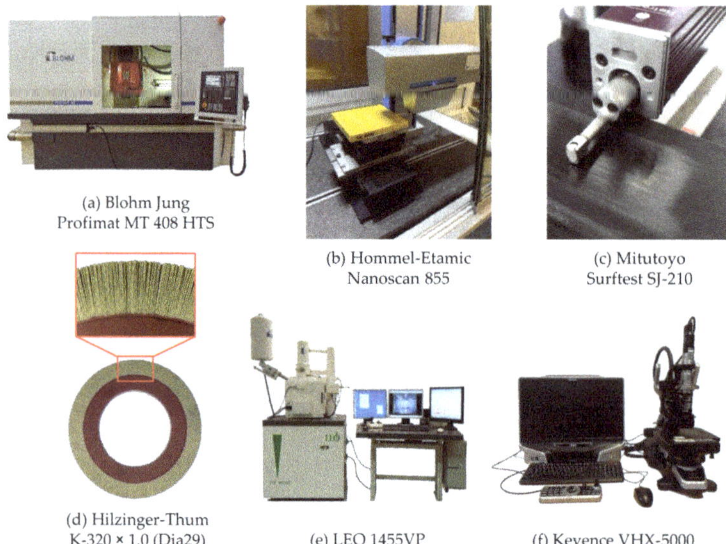

(a) Blohm Jung
Profimat MT 408 HTS

(b) Hommel-Etamic
Nanoscan 855

(c) Mitutoyo
Surftest SJ-210

(d) Hilzinger-Thum
K-320 × 1,0 (Dia29)

(e) LEO 1455VP

(f) Keyence VHX-5000

Figure 3. Experimental equipment: (**a**) profile grinding machine; (**b**) tactile surface measurement device; (**c**) tactile roughness measurement device; (**d**) exemplary circular brush used during technological investigations; (**e**) scanning electron microscope; (**f**) light microscope.

Table 1. Variation of brushing tool specification and process parameters during the technological investigations.

Grain Size s_g	Filament Diameter d_f	Brushing Velocity v_b	Feed Velocity v_{ft}	Penetration Depth a_e
mesh	mm	m/s	mm/min	mm
80	0.6	10	200 *	1 *
240	1 *	20 *	500	2
320 *		30	1000	3

* Standard tool specification or process parameter.

Before and after individual brushing cycles, the workpiece profile depth h was measured orthogonally to the brushing direction with a laser triangulator of the type optoNCDT ILD2300-2 by the manufacturer Micro-Epsilon Messtechnik Gmbh & Co. KG, Ortenburg, Germany. The sensor had a measurement range of 2 mm and was operated at a measurement frequency of f_m = 1.5 kHz and a measurement feed velocity of v_m = 100 mm/min occurring orthogonally to the brushing direction. The vertical resolution of the sensor was denoted by the manufacturer as 0.03 μm at a reference measurement frequency of f_m = 20 kHz and should analogously be smaller for lower measurement frequencies f_m. The manual fitting and subtraction of the pre-brushing profile h_0 from the post-brushing profile h_1 provided information on the material removal depth h_r. The workpiece surface roughness was determined with a tactile instrument of the type Surftest SJ-210, manufactured by Mitutoyo Corporation, Sakado, Japan (Figure 3c). Furthermore, the workpiece topography was measured with a tactile surface measurement device of the type Nanoscan 855, manufactured by Hommel-Etamic GmbH, Villingen-Schwenningen, Germany (Figure 3b). Additionally, light microscopies were made with a microscope of the type VHX 5000, manufactured by Keyence Deutschland Gmbh, Neu-Isenburg, Germany (Figure 3f). The images were taken with a 500× magnification, with a high dynamic range (HDR) enabled, and in 3D mode in order to increase the depth of focus. Scanning electron microscopy (SEM) images were obtained with a microscope of type LEO 1455VP, manufactured by Carl Zeiss

AG, Oberkochen, Germany (Figure 3e). For the evaluation of the material removal depth h_r and the workpiece topography, the software MATLAB R2019b by The Mathworks INC., Natick, MA, USA was used.

3. Results

3.1. Ground Surface Characteristics

Characterizing the surface of the ground workpieces on the basis of SEM images (Figure 4), two prominent features became apparent: Firstly, the unidirectional grooves of the preceding plane grinding process stood out, discernable as parallel lines from left to right, and the breadth and occurrence of which depends predominantly on grain size and grain distribution in the grinding tool. Secondly, the plane ground surface was scarred with unevenly spread pockmarks of various sizes and shapes. Their approximate depths could not be determined based on SEM images but are suspected to be larger than those of the unidirectional grinding grooves. Equally unclear is whether the pockmarks were caused solely by the plane grinding process or originated from the workpiece composition and sintering. The comparison with workpieces from the same batch, which were lapped instead of plane ground, showed similar pockmarks and therefore suggests that they might be induced prior to the grinding process and then exposed by it, likely due to inhomogeneous crystal formation.

workpiece:
plate, MgO-PSZ (ZrO_2), 200 × 200 × 20 mm³, plane ground

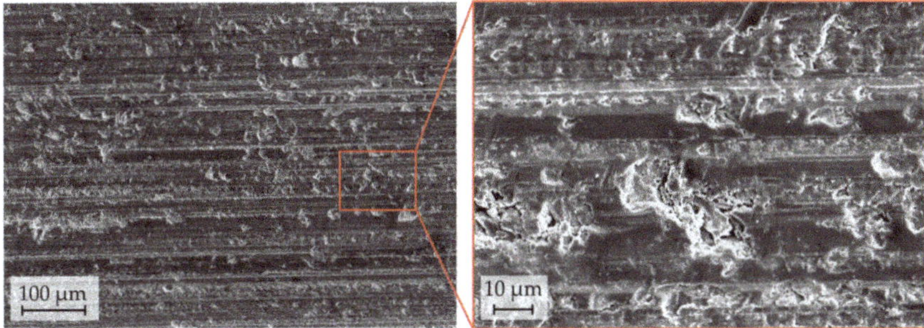

Figure 4. SEM images of the plane ground ZrO_2 surface.

The characteristic unidirectional grinding grooves and nonuniform pockmarks can also be observed on the light microscopy images (Figure 5a). The comparison with a gently brushed surface showed that both features can be partially removed by the abrasive filaments and replaced with multidirectional brushing grooves, which were thinner and shallower than the initial grinding grooves (Figure 5b).

As surface defects such as grooves and pockmarks are prone to form fractures and thus weaken the material, especially in brittle materials such as ceramics [4], their removal can increase the component durability. Considering medical engineering as the most important field of application for ZrO_2, pockmarks and other pitted blemishes in artificial dentures may lead to perpetual deposits of food debris and hence should be avoided. Similarly, rough surfaces possess a larger surface area and therefore increase bacteria accumulation [14].

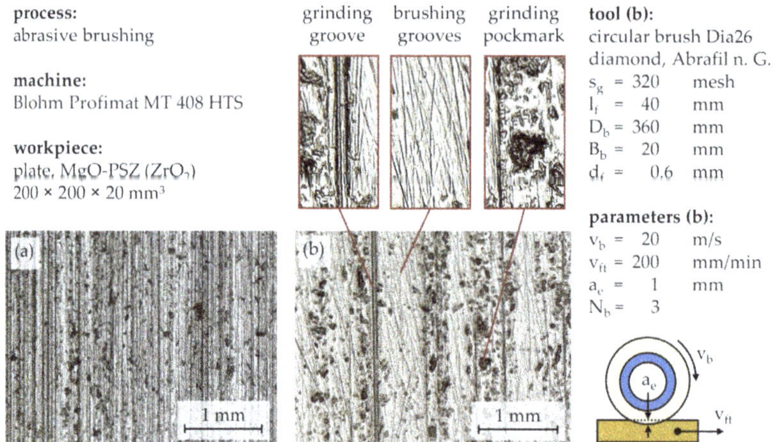

Figure 5. Light microscopy images of ZrO_2 with characteristic surface defects: (**a**) ground surface; (**b**) brushed with abrasive filaments with a filament diameter of $d_f = 0.6$ mm.

3.2. Tool Specification

Tool specification parameters, which influence the processing result, are mainly the grain size s_g, the filament diameter d_f, and the filament length l_f. While large grain sizes s_g generally lead to higher material removal rates, lower surface roughness can be achieved with small grain sizes s_g [12,15]. Therefore, the processing results for three different grain sizes s_g were compared: $s_g = 320$ mesh (29.2 µm), $s_g = 240$ mesh (44.5 µm), and $s_g = 80$ mesh (185 µm) [16,17] (Figure 6). Operated at standard process parameters, after a number of brushing cycles $N_b = 3$, the processing results of the tools with relatively fine grain sizes of $s_g = 320$ mesh and $s_g = 240$ mesh appeared similar, with unidirectional grinding grooves replaced to some extent by multidirectional brushing grooves and pockmarks being exposed rather than removed. However, the comparison with a coarse grain size of $s_g = 80$ mesh showed that all unidirectional grinding grooves and most pockmarks were replaced by multidirectional brushing grooves, which were wider and seemed deeper than the brushing grooves created with smaller grain sizes s_g.

As light microscopy images are only suitable for a qualitative analysis of the processing result, the workpiece topography was furthermore measured with a tactile instrument before and after brushing. Due to the assumed axial symmetry of the brushed profiles, the measurement tip was placed in a reproducible position near the center of each brushed profile and moved outwards in axial direction x_m. The measured surface comprised a projected area with a length in the axial direction of $x_m = 10$ mm and a width in the feed direction of $y_m = 4$ mm. This led to a total of 41 roughness profiles per brushed surface and a 0.1 mm distance between single roughness profiles.

Considering the topographies before and after brushing with a grain size of $s_g = 320$ mesh and standard process parameters (Figure 7), a preservation of the workpiece waviness can be observed alongside with an alteration of the surface roughness, characterized by the retaining of roughness valleys and the removal of roughness peaks. This suggested that successive abrasive brushing with fine grains subjected to a force-controlled principle allowed for the surface finishing of ZrO_2 without the formation of entirely new surfaces, as would be the case for similar finishing processes such as fine grinding or lapping, subjected to position-controlled principles.

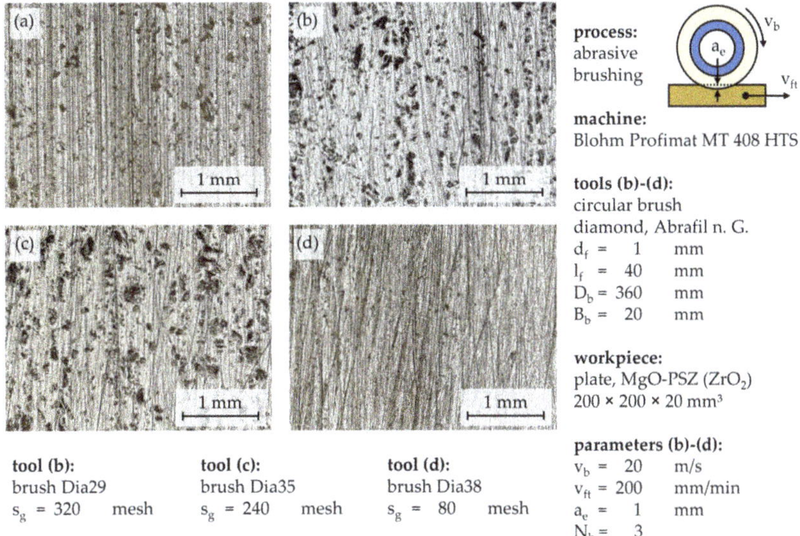

Figure 6. Light microscopy images of ZrO_2 under the variation of the grain size s_g: (**a**) ground surface; (**b**) brushed with a grain size of s_g = 320 mesh; (**c**) brushed with a grain size of s_g = 240 mesh; (**d**) brushed with a grain size of s_g = 80 mesh.

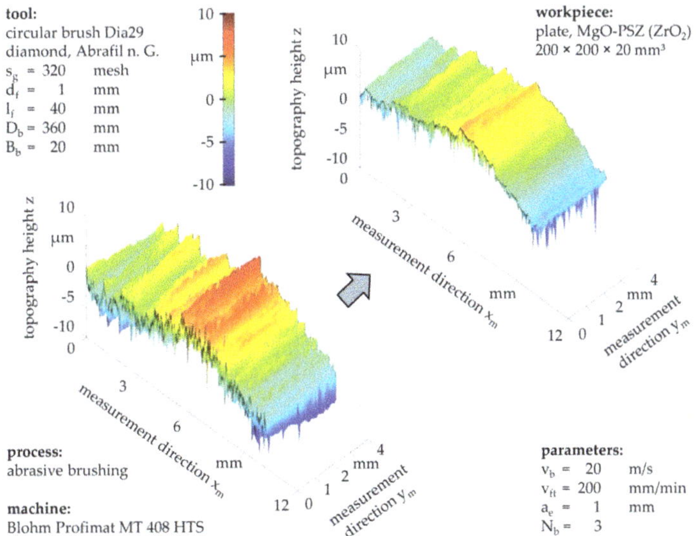

Figure 7. Topography of ZrO_2, brushed with a grain size of s_g = 320 mesh.

The comparison between the topography of a workpiece brushed with fine grains with a size of s_g = 320 mesh (Figure 7) and the one which was also brushed with standard process parameters but with coarse grains with a size of s_g = 80 mesh (Figure 8) suggested that the material removal mechanisms strongly depended on the grain size s_g.

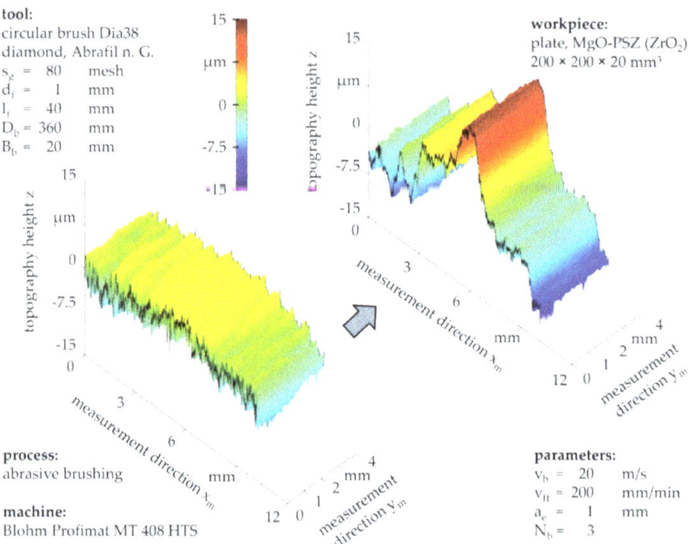

Figure 8. Topography of ZrO$_2$, brushed with a grain size of s_g = 80 mesh.

Whereas fine grains altered the workpiece topography only slightly, coarse grains led to a large form deviation, exceeding the total height of the initial roughness profile and forming a new topography with arbitrary-form peaks and valleys. Depending on the application, this might affect the workpiece quality and functionality more than the presence of pockmarks, which remained yet to be detected based on tactile workpiece topography measurements.

Apart from the grain size s_g, tool specification parameters worth investigating during future research are the filament diameter d_f and the filament length l_f, both of which have an effect on the contact pressure p_c (i.e., short and thick filaments leading to high contact normal forces F_n), thus impacting the grain penetration depth and consequently the processing result [7,9]. Additional tool specification parameters with minor influence on the processing result are the grain mass percentage c_g and the type of plastic matrix used for the abrasive filaments [9]. Based on prior experimental results, finishing ceramic workpieces with grain types softer than PCD is possible but might cause excessive tool wear due to the similar hardness of abrasive grains and workpiece material.

3.3. Process Parameters

Apart from the grain size s_g, the process parameters brushing velocity v_b, feed velocity v_{ft}, penetration depth a_e, and number of brushing cycles N_b influenced the processing result, the productivity, and the tool wear. The processing parameter mostly responsible for the number of contacts between the workpiece and the cutting edges of the abrasive grains was the brushing velocity v_b, measured at the filament tip, being the outermost point of the circular brush tool. Its variation showed a significant impact on the removal of the unidirectional grinding grooves and the pockmarks (Figure 9). Whereas a brushing velocity of v_b = 10 m/s left most of the initial topography unchanged even after a number of brushing cycles of N_b = 3 (Figure 9b), most unidirectional grinding grooves and some pockmarks were removed with a brushing velocity of v_b = 20 m/s (Figure 9c), while all grinding grooves and most pockmarks were replaced by multidirectional brushing grooves with a brushing velocity of v_b = 30 m/s (Figure 9d).

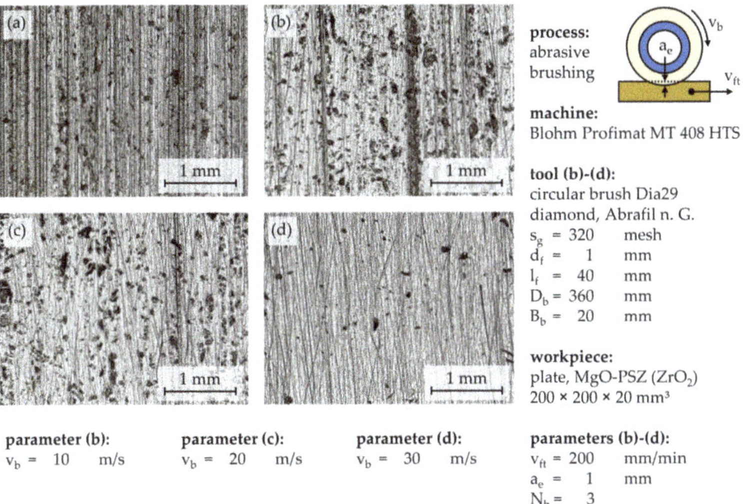

Figure 9. Light microscopy images of ZrO_2 under the variation of the brushing velocity v_b: (**a**) ground surface; (**b**) brushed with a brushing velocity of v_b = 10 m/s; (**c**) brushed with a brushing velocity of v_b = 20 m/s; (**d**) brushed with a brushing velocity of v_b = 30 m/s.

The light microscopy images indicate that the pockmarks were reduced in size as more workpiece material was removed, suggesting that the phenomenon only occurred close to the workpiece surface. This contributes to the assumption that the surface defects were caused by the initial grinding treatment.

Examining the workpiece surface after a brushing process with a brushing velocity of v_b = 30 m/s on the basis of SEM images (Figure 10), the multidirectional brushing grooves were less conspicuous than in the corresponding light microscopy images, while the complex shape of the pockmarks as well as their relatively large depths were most salient. Additionally, the edges around the pockmarks appeared rounded, although their rough inward texture depicted the brittleness of the workpiece material.

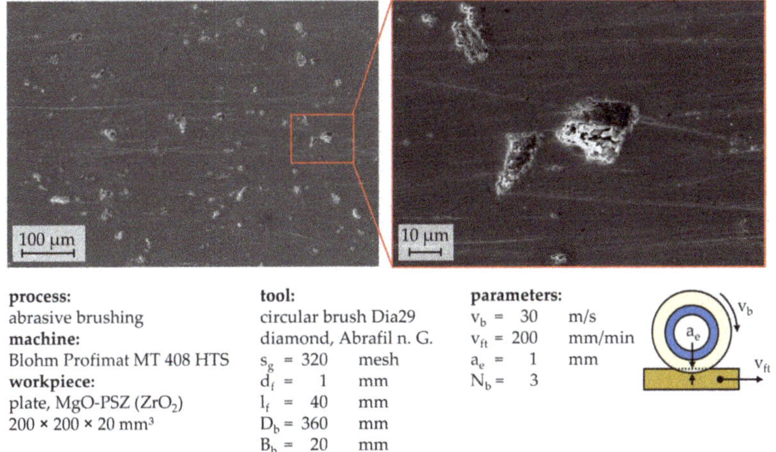

Figure 10. SEM images of ZrO_2 surface, brushed with a brushing velocity of v_b = 30 m/s.

Comparing the workpiece topographies before and after a high velocity brushing process (Figure 11), the workpiece form and waviness remained nearly the same, although all previously produced roughness features were removed, indicating the fabrication of an entirely new surface as well as a material removal depth h_r of >9.9 µm, the total height of the initial roughness profile Rt (Figure 2). The corresponding total height of the roughness profile after N_b = 3 brushing cycles was Rt = 1.9 µm, which amounted to a percental roughness reduction of ΔRt = 80.7%, approximately the same as the percental peak height reduction of ΔRpk = 84.2% and the percental valley depth reduction of ΔRvk = 82.3%. This means that roughness peaks and valleys were equally reduced.

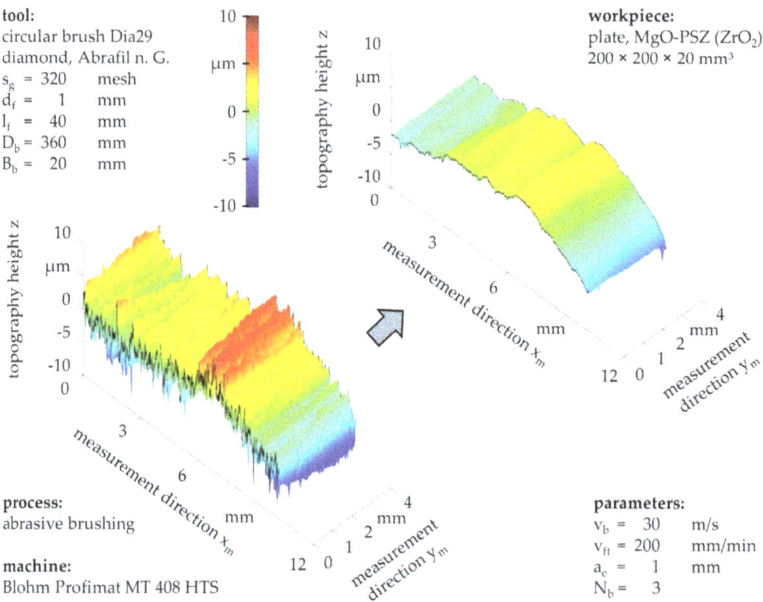

Figure 11. Topography of ZrO$_2$, brushed with a brushing velocity of v_b = 30 m/s.

In comparison, the workpiece topography after a brushing process with a brushing velocity of v_b = 10 m/s shows that only the roughness peaks were partially removed while the roughness valleys and the workpiece form remained mostly intact (Figure 12). With an initial total height of the roughness profile of Rt = 7.8 µm, the resulting total height after N_b = 3 brushing cycles was Rt = 5.5 µm, amounting to a roughness reduction of only ΔRt = 29.7%. The considerable difference between the percental peak height reduction of ΔRpk = 55.3% and the percental valley depth reduction of ΔRvk = 32.3% supported the visual observation that roughness peaks were removed more efficiently than roughness valleys. This can be attributed to the fewer and less impactful contacts between abrasive filaments and workpiece compared to processes with high brushing velocities v_b (Figures 7 and 11).

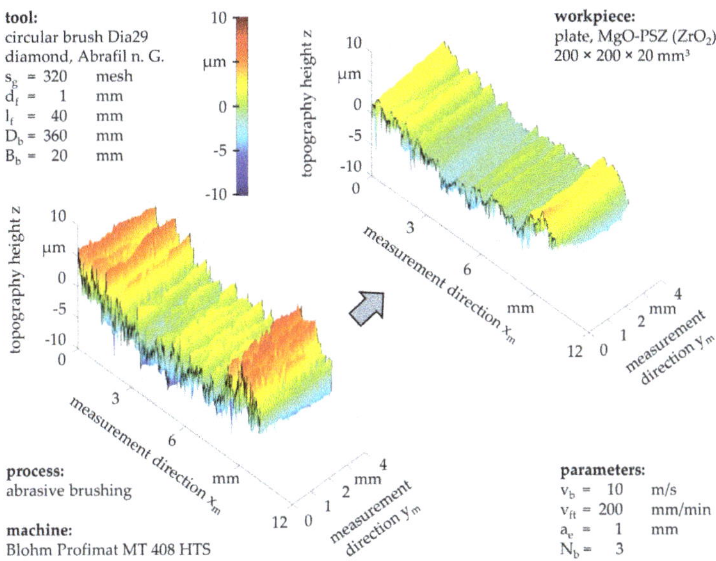

Figure 12. Topography of ZrO_2, brushed with a brushing velocity of $v_b = 10$ m/s.

In addition, the technological investigations conducted for this work confirmed an independence of the processing result from the feed velocity v_{ft}, meaning that multiple brushing cycles N_b at high feed velocities v_{ft} amounted to similar results to few brushing cycles N_b at low feed velocities v_{ft}, as long as the overall contact duration between tool and workpiece remained the same. Nonetheless, the low heat conductivity of ZrO_2 might make brushing at high feed velocities v_{ft} more feasible and reduce tool wear due to a better distributed heat flow and more cooling time between brushing cycles, which prevents the plastic matrix of the abrasive filaments from melting and leaving residue on the workpiece, avoiding additional cleaning processes.

Further technological investigations showed no apparent qualitative differences between the penetration depths of $a_e = 1$ mm, $a_e = 2$ mm, and $a_e = 3$ mm in regard to the processing result. However, contemporary research on steel workpieces (16MnCr5) indicates that large penetration depths a_e slightly increase the material removal rate, the reduction of the surface roughness, and the tool wear due to increased contact forces [7,9].

The variation of the number of brushing cycles N_b supports the established theory that the surface roughness is reduced successively and regressively, until a lower roughness limit is reached—its value depending mainly on the tool specification—after which consecutive brushing cycles N_b always yield new topographies as opposed to merely reducing certain roughness features such as roughness peaks [9].

3.4. Processing Result Deviations

Since most aggressive brushing processes, specifically those with a large grain size s_g and a high brushing velocity v_b, do not only alter the surface roughness, but may also lead to a potentially unwanted change of the workpiece form, a more detailed analysis of the proceeding shape deviation is required. Therefore, the measurement length of the workpiece topography was increased to $s_m = 50$ mm in order to capture the entire width of the brushed profile (Figure 13), as it was not homogenous due to the diverse interactions between abrasive filaments, depending highly on their positions on the brush body. The technological investigations showed that besides large grain sizes s_g and high brushing velocities v_b, a large number of brushing cycles N_b can also lead to a form deviation after the initial roughness features were completely removed. This resulted in a characteristic "W"-shape, the depth of which was successively increased with each additional brushing cycle. Likewise, noteworthy is that the width of the "W"-shape amounted to $\Delta x_m = 27.8$ mm, more than the nominal tool width of $B_b = 20$ mm, as the abrasive filaments were parted in the middle and then deflected towards both sides in the axial direction (measurement direction x_m), dodging the high pressure caused by filament interactions and taking the path of least resistance, meaning the axial side obstructed by fewer neighbor filaments. Although a sharp transition from the brushed profile to the unprocessed workpiece surface can be recognized, a small number of protruding abrasive filaments appeared to have been in contact with the unprocessed surface, as the roughness beyond the transition edge was slightly reduced.

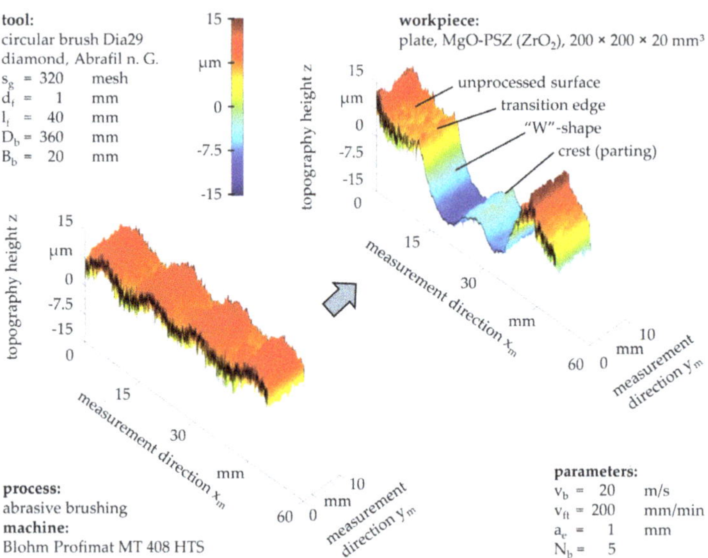

Figure 13. Topography of ZrO_2, measured across the width of an overbrushed profile.

In addition to the workpiece topography, laser distance measurements were made between brushing cycles in order to monitor the workpiece form deviation over time, characterized by the material removal depth h_r across the measurement width s_m (Figure 14). The hypothesis that the crest of the "W"-shape was formed by the parting of the abrasive filaments as opposed to being predetermined by the initial workpiece waviness, as seen in Figure 13, was supported by the fact that it was first observed after several brushing cycles, whereas the material removal depth h_r after $N_b = 1$ brushing cycles resembled a crestless "U"-shape instead. Moreover, the "W"-shapes of most brushing processes showed an asymmetry towards one axial side with less material being removed on the opposite side, presumably resulting from a minor unevenness or inclination of the brushing tool profile and/or the workpiece. As the formation of a "W"-shape was also observed on the brushing tool profile, mutually reinforcing form deviation mechanisms between tool and workpiece were most likely to occur. The technological investigations furthermore suggested that the workpiece material removed per brushing cycle followed a degressive trend, implying that rough surfaces were processed more efficiently than smooth surfaces and large numbers of brushing cycles N_b led to unproductive machining.

Apart from the aforementioned profile deviation, brushing tools are prone to lose sharpness over time due to an adjustment of the initially cylindrical abrasive filament tips to the workpiece surface, gradually transforming a sharp point contact into a flat, elliptical contact geometry [11]. Although the interrelations between wear and machining characteristics were not yet scientifically understood, first conclusions can be indirectly drawn from the processing result. One such correlation was the percental reduction of the workpiece roughness, specified by the arithmetic mean deviation of the roughness profile Ra, measured over the entire period of application t_u of the brushing tool (Figure 15).

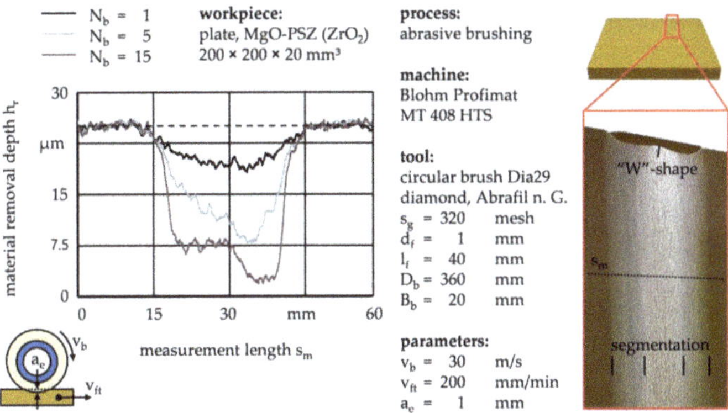

Figure 14. Material removal depth h_r, measured across the width of an overbrushed profile.

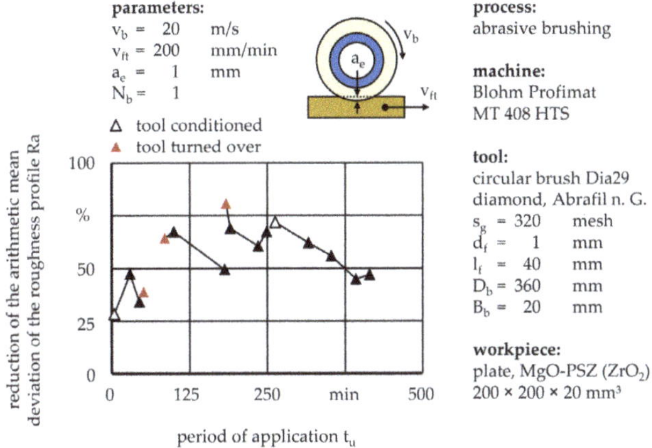

Figure 15. Deviating roughness reduction caused by the wear of the brushing tool.

After a brief initialization period, the roughness reduction continuously decreased. For the purpose of productivity, the brushing tool either needed to be expensively reconditioned or at moderate cost turned over to utilize the sharp sides of the abrasive filament tips as opposed to the flattened sides. During the technological investigations, these methods resulted in percental roughness reductions between ΔRa = 28.3% and ΔRa = 80.6% after single brushing cycles with standard process parameters, using one brushing tool with a grain size of s_g = 320 mesh. However, depending on the application, a sudden increase of the material removal rate may be undesirable, as it might cause further form deviation of the workpiece instead of merely roughness reduction. In this case, frequently turning over the brushing tool or changing the rotational direction is advised in order to maintain a consistent level of productivity, represented by the material removal rate.

4. Summary and Discussion

Brushing with bonded abrasives is a finishing process which can be used to enhance the quality of technical surfaces, primarily by decreasing the surface roughness and removing near-surface defects. This could considerably improve the machining results of ceramic materials with defects such as grooves or pockmarks caused by grinding processes, although the finishing of sintered ceramics without preceding grinding treatments is also conceivable in order to decrease the substantial machining costs [4].

Within the scope of this work, the improvement of the surface quality was demonstrated on the basis of MgO-PSZ (ZrO_2) finished with circular brushing tools, the abrasive filaments of which contained PCD. The findings showed that abrasive brushing can furthermore be utilized to create specialized surfaces by only removing roughness peaks while leaving roughness valleys intact. Abrasive filaments with small grain sizes s_g proved appropriate for this task, as large grain sizes s_g rapidly led to form deviations larger than the total height of the roughness profile Rt and the resulting coarse brushing grooves negatively affected the minimum achievable roughness. Since small grain sizes s_g usually require multiple brushing cycles due to the low material removal rate, performing one brushing cycle with coarse grains and one with fine grains might increase productivity, which remains to be confirmed by further experiments. In addition, pending are the influences of filament diameter d_f and filament length l_f on the processing result while brushing ceramics. First experiments suggested that abrasive filaments shorter than l_f = 40 mm are prone to take permanent thermal damage due to the low heat conductivity of ZrO_2, which in industrial applications could be compensated by adding cooling lubricants.

Additionally, the brushing velocity v_b was confirmed to have a significant impact on the machining characteristics, as a high brushing velocity of v_b = 30 m/s proved advantageous for the reduction of near-surface defects and yielded topographies with equally reduced roughness peaks and valleys, indicated by a percental peak height reduction of ΔRpk = 84.2% and a percental valley depth reduction of ΔRvk = 82.3%. Nevertheless, the complete removal of pre-existing roughness features might be undesirable and also cause excessive tool wear due to increased heat input. In comparison, a low brushing velocity of v_b = 10 m/s yielded a percental peak height reduction of ΔRpk = 55.3% and a percental valley depth reduction of ΔRvk = 32.3%, and their difference indicated that roughness peaks were removed more efficiently than roughness valleys.

Generally, workpiece form deviations and tool wear developed more rapidly than during comparable technological investigations priorly carried out on steel 16MnCr5 workpieces [9]. Even with fine grains and medium brushing velocities v_b, after a small number of brushing cycles N_b, the forming of a "W"-shaped workpiece profile was observed, caused by the quasi-symmetrical deflection of the abrasive filaments attached to the lateral area of the cylindrical brush body. However, this might be of no consequence, if pot-shaped brushing tools were used instead, which comprised abrasive filaments attached to the base area of the cylindrical brush body, therefore yielding different kinematic relations as well as circular brushing groove patterns on the workpiece surface.

Tool wear was measured based on the consistency of processing results, specifically the arithmetic mean deviation of the roughness profile Ra, and could be compensated by frequent turning of the brushing tool or alternating of the rotational direction. A wear study based on the change in form of single abrasive filament tips remains yet to be conducted.

Author Contributions: Conceptualization, A.H. and E.U.; methodology, E.U.; software, A.H.; validation, A.H.; formal analysis, A.H.; investigation, A.H.; resources, A.H.; data curation, A.H.; writing of original draft preparation, A.H.; writing of review and editing, E.U.; visualization, A.H.; supervision, E.U.; project coordination, E.U.; funding acquisition, E.U. All authors have read and agreed to the published version of the manuscript.

Funding: This research was funded by Deutsche Forschungsgemeinschaft (DFG) within the scope of the project "Analyse des Zerspan- und Verschleißverhaltens beim Bürstspanen mit abrasivem Medium sprödharter Werkstoffe" (project number: 392312434).

Acknowledgments: The authors kindly thank the funder for their support.

References

1. Uhlmann, E.; Lypovka, P.; Sommerfeld, C.; Bäcker, C.; Dethlefs, A.; Hochschild, L. Abrasives Bürsten. *Werkstatt Betrieb (WB)* **2014**, *WB 4/2014*, 70–72.
2. Hao, L.; Lawrence, J.; Chian, K.S. Osteoblast cell adhesion on a laser modified zirconia based bioceramic. *JMSMM* **2005**, *16*, 719–726. [CrossRef] [PubMed]
3. Forkas-Tsentzeratos, G. Influence of the Surface and Heat Treatment on the Flexural Strength and Reliability of Y-TZP Dental Ceramic. Ph.D. Thesis, Medicinal Faculty of the Eberhard Karls University, Tübingen, Germany, 2010.
4. Birkby, I.; Stevens, R. Applications of Zirconia Ceramics. *Key Eng. Mater.* **1996**, *122–124*, 527–552. [CrossRef]
5. Emami, M.; Sadeghi, M.H.; Sarhan, A. Minimum Quantity Lubrication in Grinding Process of Zirconia (ZrO2) Engineering Ceramic. *IJMMME* **2013**, *1*, 187–190.
6. Fook, P.; Riemer, O. Characterization of Zirconia-Based Ceramics After Microgrinding. *JMNM* **2019**, *7*. [CrossRef]
7. Uhlmann, E.; Sommerfeld, C.; Renner, M.; Baumann, M. Bürstspanen von Profilen. *wt Werkstattstechnik Online* **2017**, *107*, 238–243.
8. Rentschler, J.; Muckenfuß, G. Neue Anwendungsmöglichkeiten durch hochtemperaturbeständige Schleiffilamente in der Oberflächenbearbeitung. In *Jahrbuch Honen, Schleifen, Läppen und Polieren*; Hoffmeister, H.W., Denkena, B., Eds.; Vulkan: Essen, Germany, 2013; pp. 387–403.
9. Uhlmann, E. Flexible Feinstbearbeitung von Funktionsflächen mit alternativem Werkzeugkonzepten (FlexFeinst). In *Schlussbericht zu IGF-Vorhaben Nr. 19601 N/1*; Technische Universität Berlin, Institut für Werkzeugmaschinen und Fabrikbetrieb: Berlin, Germany, 2020.
10. Przyklenk, K. *Bestimmung des Bürstenverhaltens anhand einer Einzelborste. Berichte aus dem Fraunhofer-Institut für Produktionstechnik und Automatisierung (IPA), Stuttgart, Fraunhofer-Institut für Arbeitswirtschaft und Organisation (IAO), Stuttgart und Institut für Industrielle Fertigung und Fabrikbetrieb der Universität Stuttgart Nr. 87*; Warnecke, H.J., Bullinger, H.-J., Eds.; Springer: Berlin/Heidelberg, Germany, 1985.
11. Overholser, R.W.; Stango, R.J.; Fournelle, R.A. Morphology of metal surface generated by nylon/abrasive filament brush. *Int. J. Mach. Tools Manuf.* **2002**, *43*, 193–202. [CrossRef]
12. Landenberger, D. Flexible Feinbearbeitung für die Refabrikation von Automobilkomponenten. Ph.D. Thesis, Universität Bayreuth, Bayreuth, Germany, 2007.
13. Hochschild, L. Finishbearbeitung Technischer Oberflächen aus Gehärtetem Stahl unter Verwendung von Rundbürsten mit Schleiffilamenten. Ph.D. Thesis, Technische Universität, Berlin, Germany, 2018.
14. Hussein, A.I.; Hajwal, S.I.; Ab-Ghani, Z. Bacterial Adhesion on Zirconia, Lithium Desilicated and Gold Crowns-In Vivo Study. *Adv. Dent. Oral Health* **2016**, *1*, 1–3. [CrossRef]
15. Landenberger, D.; Steinhilper, R.; Rosemann, B. Verbesserung der Oberflächengüte durch Bürstspanen. *VDI-Z* **2007**, *149*, 67–69.
16. Federation of European Producers of Abrasives. *FEPA-Standard 42-1: Grains of Fused Aluminium Oxide, Silicon Carbide and Other Abrasive Materials for Bonded Abrasives and for General Applications Macrogrits F4 to F220*; Federation of European Producers of Abrasives: Paris, France, 2006.
17. Federation of European Producers of Abrasives. *FEPA-Standard 42-2: Grains of Fused Aluminium Oxide, Silicon Carbide and Other Abrasive Materials for Bonded Abrasives and for General Applications Microgrits F230 to F2000*; Federation of European Producers of Abrasives: Paris, France, 2006.

Publisher's Note: MDPI stays neutral with regard to jurisdictional claims in published maps and institutional affiliations.

Article

The Proposition of an Automated Honing Cell with Advanced Monitoring

Adam Barylski and Piotr Sender *

Department of Manufacturing and Production Engineering, Gdańsk University of Technology,
80-233 Gdańsk, Poland; abarylsk@pg.edu.pl
* Correspondence: piotr.sender@pg.edu.pl; Tel.: +48-662-082-579

Received: 3 September 2020; Accepted: 22 October 2020; Published: 28 October 2020

Abstract: Honing of holes allows for small shape deviation and a low value of a roughness profile parameter, e.g., Ra parameter. The honing process heats the workpiece and raises its temperature. The increase in temperature causes thermal deformations of the honed holes. The article proposes the construction of a honing cell, containing in addition to CNC honing machine: thermographic camera, sound intensity meter, and software for collecting and analyzing data received during machining. It was proposed that the level of sound intensity obtained during honing could be monitored continuously and that the images from a thermographic camera could be analyzed on-line. These analyses would be aimed at supervising honing along with the on-line correction of machining parameters. In addition to the oil cooler, the machining cell may have an automatic selection of the grain trajectory shape, with specified value of the radii of curvature of the abrasive grain trajectories, according to the wall thickness of the honed workpiece, which will result in reducing the temperature generated during honing. Automated honing cell can mostly increase honing process efficiency. Simulations in FlexSim showed the possibility of increasing the efficiency of the honing process more than 20 times.

Keywords: honing of holes; kinematics of honing; automation of honing; abrasive grain trajectories; FlexSim; thermography

1. Introduction

The honing process heats the workpiece and raises its temperature. Specialist literature dealing with the honing process of cylindrical surfaces draws attention to the issue of the temperature increase during machining [1–11], which causes the thermal deformation of the honed hole. The temperature increase is also a very important issue during the honing process of flat surfaces by grinding with lapping kinematics as shown in [12]. Another interesting issue, in addition to the coexisting issue, i.e., surface texturing [13], is the use of new honing methods [14–17] and honing equipment used during machining [18,19] to improve the quality of the holes obtained [20–24].

The process of honing is influenced by the selection of proper machining parameters [1,25–30], which directly affects the obtained cylindricity deviations of the honed holes [9,10,31–35]. Much attention in the literature has also been devoted to examining the possibility of creating a new honing equipment and machines [18] as well as machining strategy, during which a different shape of oil channels cross-hatch could be obtained [12,36–46] which influences the further development of honing methods [47–52].

Irene Buj-Corral [2] described the methodology of selecting the appropriate density of the abrasive whetstone based on the analysis of the acoustic signal recorded in honing process and checked that roughness of honed cylinders increases mainly with abrasive grain size, followed by honing head pressure. She researched that tangential speed influences roughness slightly [53]. Similarly, Chavan [38] describes the effect of the pressure of the abrasive whetstone and the honing speed on the obtained roughness profile parameters. Deepak [27] dealt with in detail the influence of honing parameters on the obtained parameters of the roughness profile of honed surface. He stated that the rotational speed of the head had the greatest influence on the R_z, R_k, and R_{vk} parameters, while the linear speed of the jump had the greatest influence on the R_{pk} parameter. Barylski [1] indicates that the selection of parameters affects the size of the honed material and affects the amount of temperature generated in the workpiece.

The latest literature describes the use of the neural network method to supervise the honing process [54], but the analysis of parameters leads to obtaining the desired roughness profile parameters. In this article, it is proposed that the neural network completely diagnose the process of honing, including the analysis of the sound signal obtained during honing and the analysis of thermograms recorded during honing, which is a novelty in proposing the direction of the honing process and its automation.

2. Kinematics of Honing Process

Goeldel [41,55] proposed to use new shapes of grain paths for honing of cylindrical holes. Bujukli [56] describes the effect of the honing head stroke length on the improvement of the cylindrical shape deviation of the honed hole.

Figure 1 shows the cross-section of the honing head and the speeds occurring in the machining system: V_{ax}—axial linear speed of the honing head [m/min], V_{az}—peripheral speed of the honing head [m/min], V_{exp}—infeed speed of abrasive whetstone [μm/min].

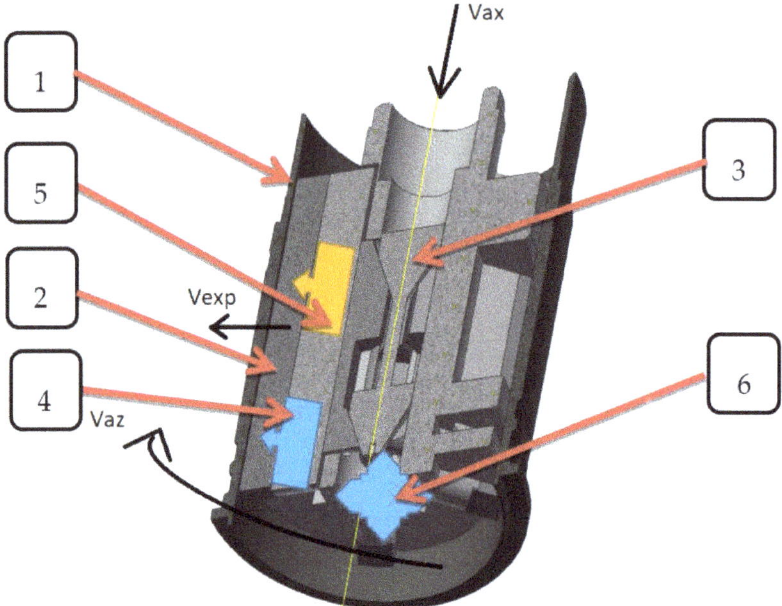

Figure 1. Honing head with main honing speeds; 1—honed cylinder liner, 2—abrasive whetstone, 3—expanding mandrel, 4—pressure and machined diameter measurement, 5—temperature measurement, 6—vision system.

The honing head has a vision system (item 6), which records the real-time image of the honed surface, introducing data to the neural network that verifies the number of oil channels produced in such a way that the continuity for the flowing oil is broken. The gauge of the pressure force and the diameter of the machined hole (item 4) and the temperature gauge (item 5) enable the supervision of the honing process in real time.

The basic formulas that define the honing kinematics (Figure 1) are:

- resultant cutting speed [m/min]

$$V_c = \sqrt{V_{ax}^2 + V_{az}^2} \tag{1}$$

- linear speed of the head in reciprocating motion [m/min]

$$V_{ax} = 2Ln_{ax} \tag{2}$$

where: L is the length of the head stroke in the axial direction, in reciprocating motion [m]; n_{ax} is the head stroke frequency in reciprocating motion [1/min]; V_{az} is the peripheral speed of the head [m/min]

$$V_{az} = \frac{\pi dn}{1000} \tag{3}$$

where: n is the head rotation speed [min^{-1}]; d is the diameter of the honed hole [mm];

And

$$tg\alpha = \frac{V_{ax}}{V_{az}} \tag{4}$$

where: α is the honing angle [°].

The honing angle is a parameter influencing the oil consumption and the amount of toxic compounds emitted to the environment during the operation of internal combustion engines [57]. It also affects the coefficient of friction of the piston rings against the cylinder surface [38,42,58,59], which has a direct impact on engine power losses.

The honing angle depends on the mutual relation of the rotational speed and the speed of the axial linear stroke of the head. A higher value of the rotational speed with a lower value of the honing head stroke speed makes the angle closer to the horizontal direction, a lower value of the rotational speed with a lower value of the head stroke speed makes the angle closer to the vertical direction.

Figure 2 shows an exemplary trajectory of the abrasive grain (item 1) obtained during honing on the developed surface of the machined hole (item 2). The advantage of the honing cell is the automatic selection of the shape of the abrasive grain trajectory, with different values of the grain trajectory curvature, which directly affects the amount of temperature generated in the workpiece.

There are a many combinations possible in the honing process of oil channel angle. This diversity is presented in Table 1. Various honing researchers and different engine manufacturers suggest the use of different honing angles, but this article proposes the use of honing with variable kinematics, which enables the creation of a scratch grid in the form of curvilinear oil channels.

Honing with traditional kinematics is carried out without a change in the machining parameters. Figure 3 shows traditional honing process, where rotational speed of honing had has a constant value during process. This honing method allows to create cross-hatch shape of abrasive grain trajectories. The rotational speed of the honing head at the beginning of the cycle increases from zero to a constant value (Figure 3a), the feed value increases from zero to a constant value (Figure 3b). In honing with constant kinematics, the stroke length increases linearly (Figure 3c), while the grain path length increases according to the curve shown in Figure 3d.

Honing with variable kinematics is carried out with a change in the machining parameters. Figure 4 shows non-traditional honing process, where rotational speed of honing had has a different rotation speed in one single cycle of honing process. This honing method allows to create different shape of abrasive grain trajectories. The rotational speed of the honing head at the beginning of the cycle increases from zero and can still increase or decrease in turn (Figure 4a), the feed value increases from zero and can increase and decrease in turn, etc., the return of the speed vector can change to the opposite in any way (Figure 4b). In honing with variable kinematics, the stroke length increases not linearly but curvilinear and according to the rotational speed and stroke speed value (Figure 4c), while the grain path length increases according to the curve shown in Figure 4d.

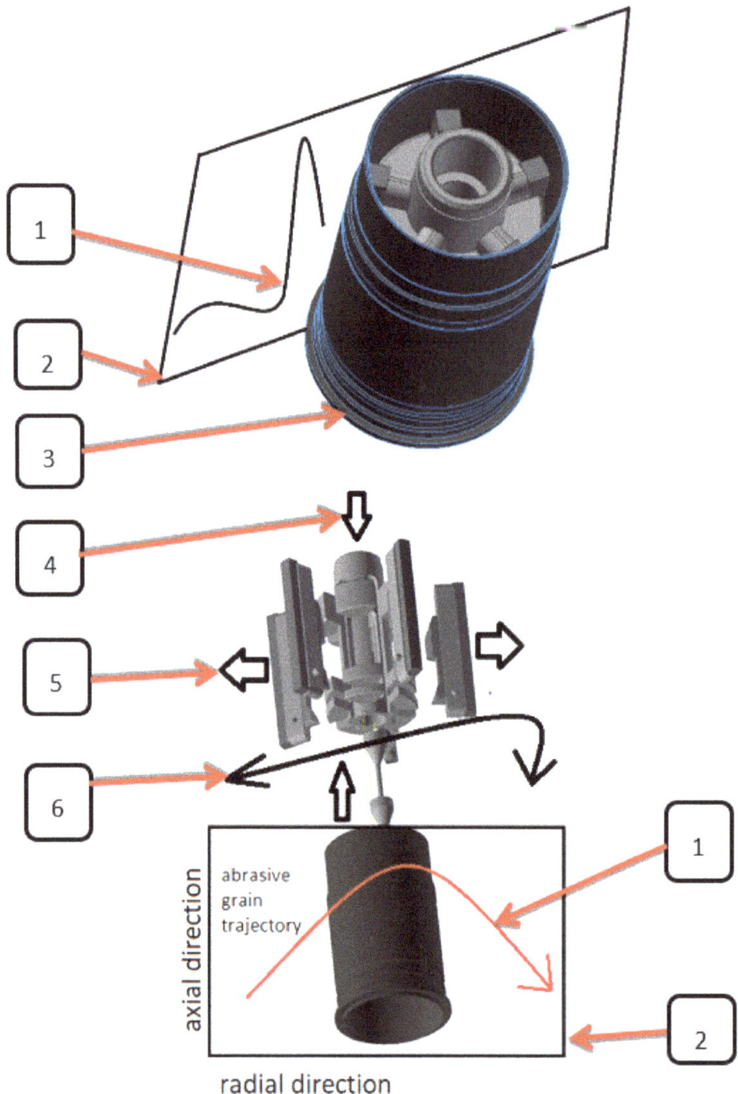

Figure 2. Possible directions of movement of honing head in honing of cylindrical holes: 1—abrasive grain trajectories, 2—expanding of honed surface, 3—honed cylinder liners with honing head, 4—additional honing head oscillation motion in vertical direction, 5—additional honing head oscillation motion in horizontal direction, 6—additional oscillation motion of honing head rotation direction.

Table 1. The honing angles discussed in the literature.

Authors or Companies	Literature Item Number	Honing Angle [°]
Bouassida H.	[57]	
Entezami S., Farahnakian M., Akbari A.,	[59]	45
Karpuschewski B., Welzel F., Risse K., Schorgel M.	[60]	
Mansori El. M., Goeldel B., Sabri L.	[61]	
Brush research manufacturing Co. Inc.	[62]	25–30
Buj-Corral I., Vivancos-Calvet J.	[3]	36.9; 38.6; 53.1
Chavan P.S., Harne M.S.	[38]	46–57
Dahlmann D., Denkena B.	[63]	110
Graboń W., Pawlus P., Wos S., Koszela W., Wieczorowski M.	[42]	15.5; 55; 72; 125
Demirci I., Mezghani S., Yousfi M., El Mansori M.	[64]	50–130
Deshpande A.K., Bhole H.A., Choudhari L.A.	[65]	25–75
Fiat Chrysler America	[66]	36
Goeldel B., Mansori M.,	[40]	
Graboń W., Pawlus P., Sep J.	[67]	
Michalski J., Woś P.	[68]	50
Reizer R., Pawlus P., Galda L., Graboń W., Dzierwa A.	[69]	
Pawlus P., Cieslak T., Mathia T.,	[52]	
Goeldel B., Mansori M.,	[55]	45, 135
Jocsak J.	[43]	30, 35, 40, 45, 60, 90
Johansson S., Nilsson Per.H., Ohlsson R., Anderberg C., Rosen B.G.	[44]	
		40, 140
Johansson S., Nilsson Per.H., Ohlsson R., Anderberg C., Bengt-Goran Rosen	[45]	
Kim J.K., Xavier F.A., Kim D.E.	[70]	
Kapoor J.,	[71]	15–22
Knoll G., Rienacker A.,	[48]	10, 30, 60, 90, 120, 150, 170
KS Motor Service International GmbH	[72]	40, 60, 80
Lawrence D.K., Ramamoorthy B.	[27]	41, 48, 51, 54, 59, 61, 64, 71, 74, 84, 88, 89, 102, 105, 111
Mezghani S., Demirci I., Yousfi M., Mansori E.M.	[49]	40–60; 120–140
Mezghani S., Demirci I., Zahouani H., Mansori E.M.	[73]	10, 20, 30, 40, 50, 60, 70, 80, 100, 110, 130, 140, 150
Obara R.B., Souza R.M.	[51]	66
Reizer R. Pawlus P.	[74]	53
Pimpalgaonkar M.H.,	[75]	20–60
Qin P.P., Yang C.I., Huang W., Xu G.W., Liu C.J.	[76]	30, 45
Sabri L., Mezghani S., Mansori E.M., Le Lan Jean-Vincent	[77]	51.14; 30–60; 140
Gashev E.A., Muratov K.R.,	[39]	
Polyanchikov Yu. N., Plotniko A.L., Polyanchikova M.Yu., Kursin O.A.,	[78]	variable angle
Sender P.	[5–8]	
Yousfi M.	[79–85]	
Tripathi B.N., Singh N.K., Vates U.K.	[20]	25–75
Yuan S., Huang W., Wang X.,	[13]	45, 90
Deepak Lawrence K., Ramamoorthy B.	[27]	43, 50, 53, 56, 60, 63, 68, 74, 79, 81, 94, 106, 108, 114
Ozdemir M., Korkmaz M.E., Guanay M.B.Buj-Corral I.,	[86]	40–80
Vivancos-Calvet J., Coba-Salcedo M.	[87]	
Q. Wang, Q. Feng, Q.F. Li and C.Z. Ren	[88]	83.70

Table 1 shows the honing angles that are discussed in the specialist literature.

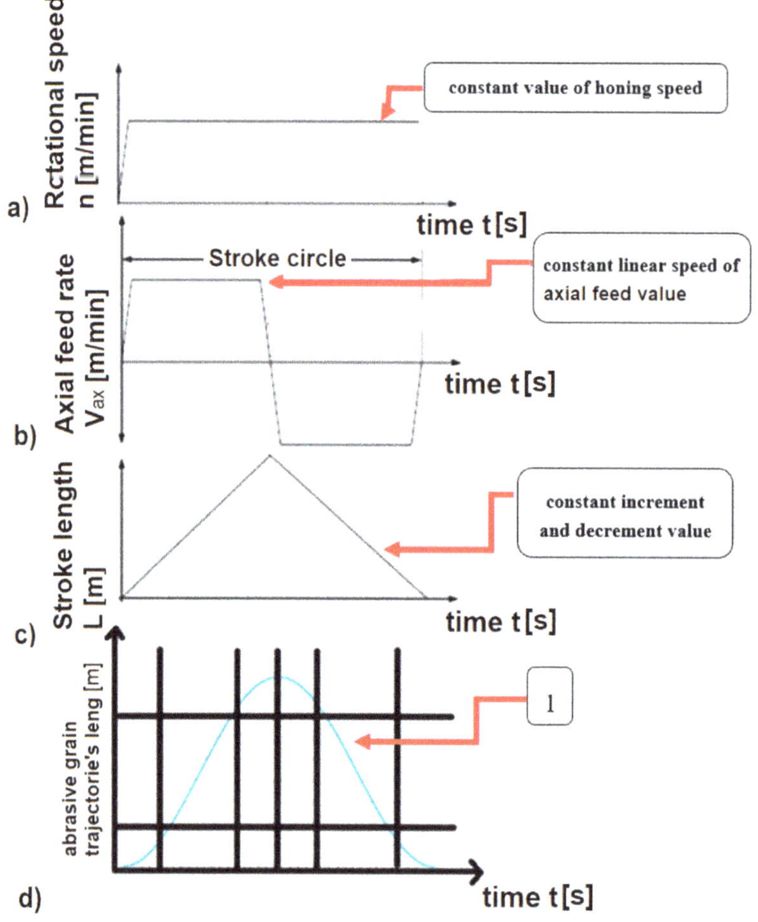

Figure 3. Traditional kinematics of honing process, (**a**) constant value of honing speed, (**b**) constant linear speed of honing head, (**c**) linear value of honing head, (**d**) length of abrasive grain path in traditional honing process; 1—an example of abrasive grain path.

Because of the disordered distribution of the grain in the abrasive stone, the grain fracture planes are at any angle to the grain direction, determined by the vector of the resultant honing speed V_c. During honing, the direction of the forces acting on the grain changes. Some of the grains are too weak in a given plane to transfer cutting forces, so that new cutting edges are constantly created during the honing process. Changing the direction of grain work also prevents the deposition of the processed material particles on the grain working surfaces; because of this phenomenon the change of the direction of the velocity vector V_c, i.e., variable honing kinematics, is an important parameter influencing the course of the process [5–8] and reduces the friction coefficient in the piston-cylinder assembly [80,82,84,85]. The value of normal acceleration a_n is responsible for the change of the vector V_c direction and the trajectory curvature.

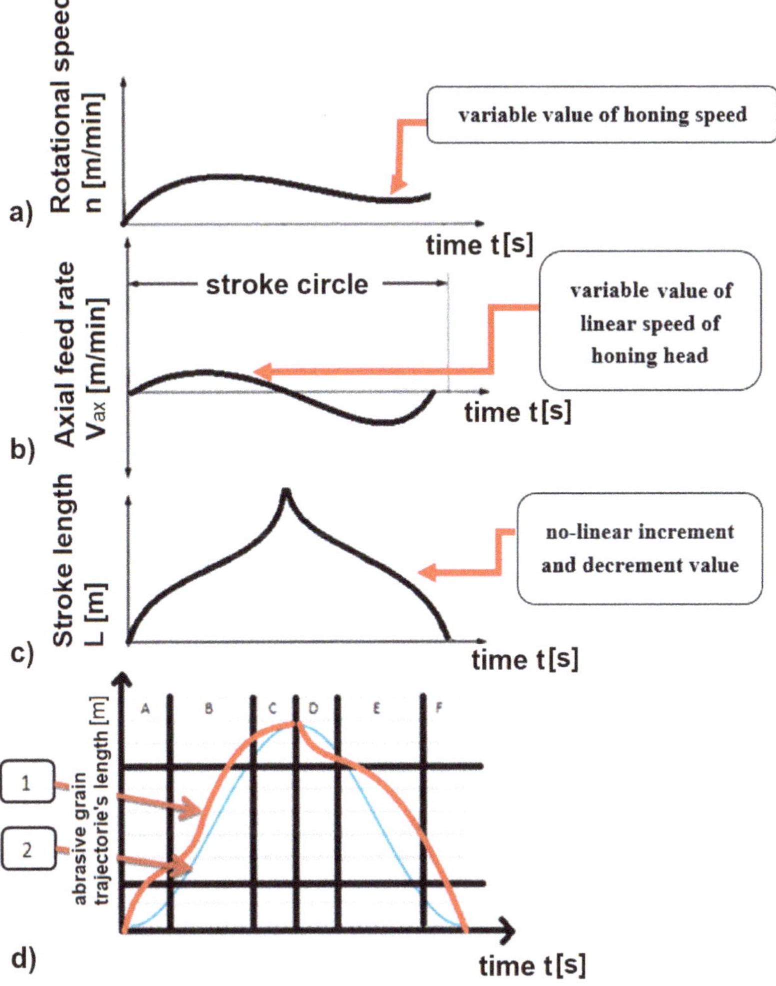

Figure 4. Variable kinematics of honing process—an example of abrasive grain trajectory obtained during honing with variable linear and tangential speed. Components of the cylindrical honing process: 1—an example of abrasive grain path received in variable kinematics of honing, 2—comparative abrasive grain path obtained in traditional honing.

3. Methods

The article proposes the construction of a honing cell, containing in addition to CNC honing machine: thermographic camera, sound intensity meter, and software for collecting and analyzing data received during machining. It was verified that the main factor hindering the serial honing treatment is the deformation of the shape of the honed hole, caused by the heating of the workpiece due to the friction of the abrasive stone against the honed surface, which also causes a change in the diameter of the hole. The prevention of this phenomenon consists in controlling the temperature to which the honed workpiece is heated during machining, and in carrying out the treatment in such a way that the amount of temperature increase is reduced due to the selection of proper honing parameters, depending on the shape and *CCR* (the curve curvature radius) of the abrasive grain trajectories [5].

Honing cell principle of operation: 1—numerical simulation of deformations, stresses, and heat flow, 2—programming of honing head movements with the selection of the appropriate shape of the abrasive grain path adjusted to the thickness of the section or sections of the honed workpieces, 3—supervising of honing during the process, measuring of diameter and cylindricity of honed hole, measuring of sound signal, 4—correcting of actual machining parameters.

3.1. The Equipment of the Honing Cell

Figure 5 shows a CNC vertical milling center equipped with the honing instrumentation, a thermographic camera, surface roughness measure gauge, and sound intensity meter. Each of the three shown laptops analyzed different signals and their output values obtained during the honing process.

The key issue of the proposed automated honing cell is to be able to supervise the honing process by analyzing the audio signal and by analyzing the images obtained from the thermographic camera during the process. The task of supervision of honing process should be to generate abrasive grain paths with different shape of grain trajectories and with different radii of curvature, depending on the data obtained from the analysis of the acoustic signal and from the analysis of the thermogram of the workpiece.

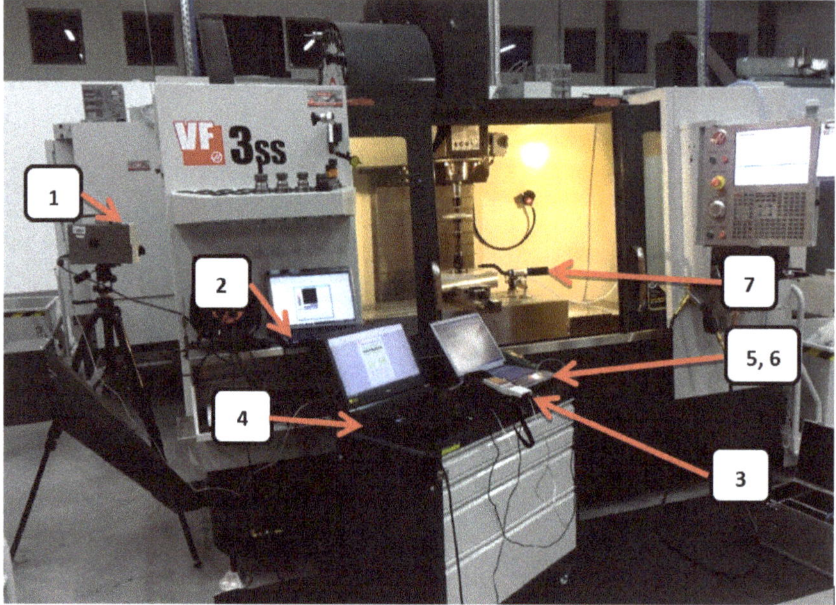

Figure 5. Haas VF-3SS milling machine with honing equipment: 1—thermal imaging camera, 2—software of thermal imaging camera, 3—Mitutoyo SJ-210 roughness meter, 4—software of Mitutoyo SJ-210 roughness meter, 5—sound intensity meter, 6—vibration meter, 7—air nozzle.

3.2. Numerical Simulations of Honing Process

The honing cell should verify the influence of honing parameters on the size of the temperature increase, which is related to thermal deformation of the honed hole. The measurement should be carried out before machining, by performing a numerical simulation of honing.

Figure 6 shows the image recorded during the numerical simulation, showing the heat flux flow through the honed cylinder liner. The occurrence of different values of heat flux is clearly noticeable, depending on the thickness of the section of honed workpiece. The differences affect the non-uniform cylindrical deformation, more information is included in [5–8]. The simulation will provide information about the size of deformation for various possible variants of machining parameters, e.g., the amount of pressure of the abrasive whetstone on the workpiece.

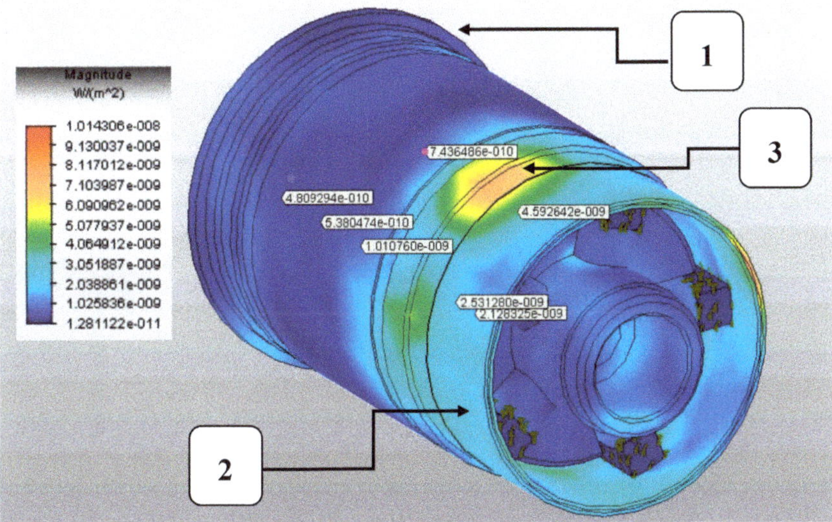

Figure 6. Image from computer simulation of honing process—heat flux flow; 1—the thickest machined cross–section, 2—the thinnest honed cross-section, 3—maximum value of the measured heat flux.

The different temperature value of the honed workpieces causes the occurrence of different values of thermal stress in different places of the workpiece and affects the deformation of the cylinder shape, which is an undesirable phenomenon.

The honing cell should verify the influence of honing parameters on the amount of stresses and deformations occurring during honing in the workpiece. The measurement should be carried out before machining, by performing a numerical simulation of honing (Figure 7).

3.3. Programming of the Grain Trajectories in Non-Conventional Way

It was verified in [5] that the shape of the abrasive grain trajectories obtained in the honing process influenced the size of the temperature rise in the honed workpieces. In addition, abrasive grain trajectories can be generated using mathematical functions such as $\sin(x)$, $\cos(x)$ etc., which would allow the creation of a path of any shape, also curves with different radii of curvature.

A very interesting issue is the problem of programming the path of the abrasive grain, different one than the traditional path resultant from the rotational and linear speeds of the honing head, which is so far used in most honing machines manufactured by leading manufacturers. In the proposed approach, a curvilinear path can be selected and generated depending on the cross-sections of the honed workpiece. It is defined in the form of a mathematical formula that defines a curvilinear path of various shapes with variable shape parameters (radius of curvature, amplitude size, frequency of change of direction). This kind of non-conventional programming can be realized in the CAD/CAM system, and previously planned in *CCR* (curve curvature radius) module.

Figure 2 shows the window view from the CAD/CAM Alphacam software, where it is possible to define a curvilinear path by adding a circular vibration to the path of varying magnitude and frequency of change of abrasive grain move direction.

Changing the shape of the abrasive grain path from a circular path to a sinusoidal path. Most basic form of sine wave describing the time function t is:

$$y = A \sin(\omega t + f) \tag{5}$$

where: A—amplitude, ω—pulsation in radians per second (closely related to the frequency in hertz), f—phase shift (if the phase is different from zero, the function graph looks shifted in time by 0 s).

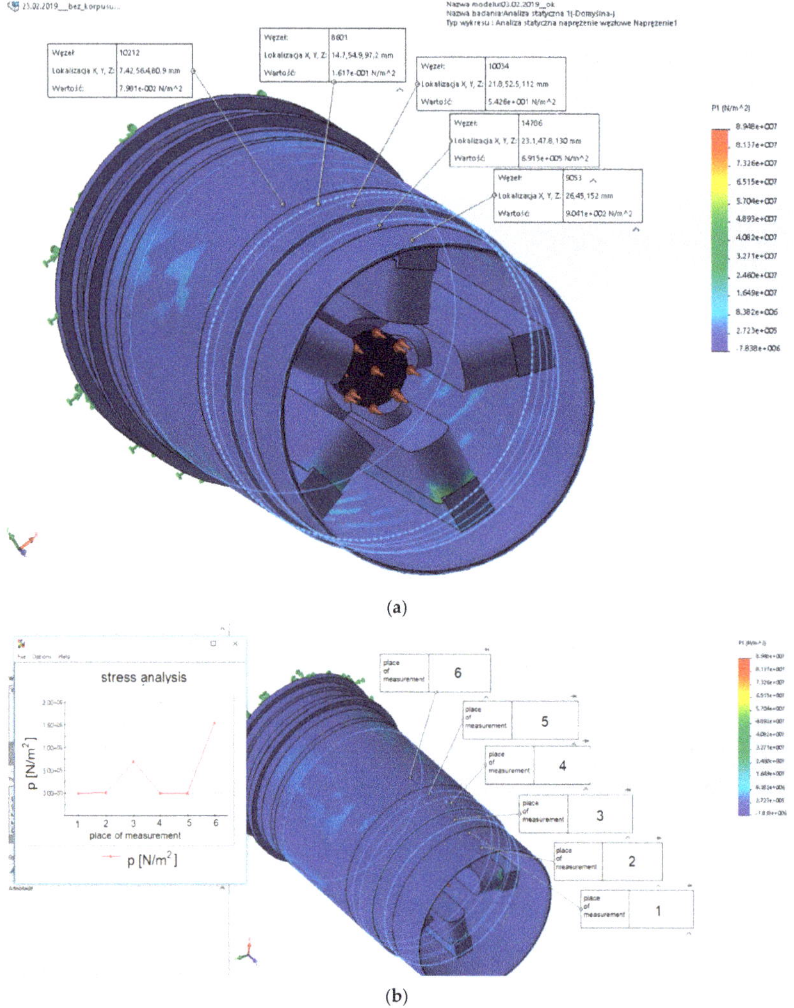

(a)

(b)

Figure 7. Deformation of the honed workpieces—different value in different places. (**a**) view of the window from the simulator with the effect of the deformations obtained, (**b**) view of the window from the simulator with the presentation of non-linear deformation results of the honed workpiece.

The curves describing the trajectory of the abrasive grain may have various shapes, characterized by a different curvature of the grain trajectory [5]. The grain trajectory may take any shape, while performing which the honing head may more or less frequently change the direction of the axial movement (Figure 8). The quality of the hole made depends on the accuracy of the honing head movement [89,90].

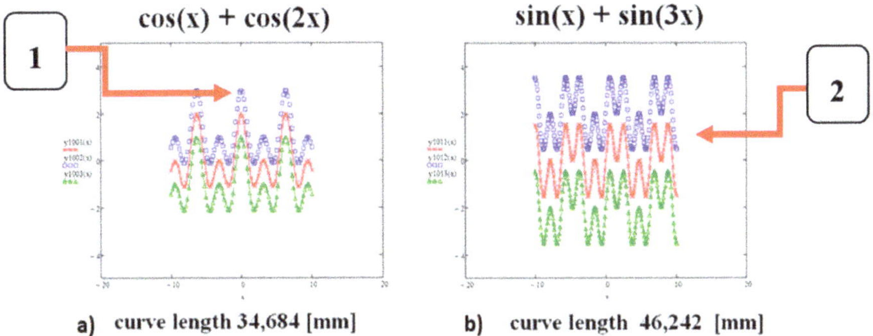

Figure 8. Examples of different grain trajectory shapes, shown on the developed surface of machined hole, with different oscillation frequency, with different path length for the same length on horizontal direction of the treated surface. The function marked with digit 1 has one change of the head movement direction in the lower and upper turning point. The function marked with digit 2 has one change in the head movement direction in the lower turning point and two changes in the direction of the upper turning point.

Figure 9 shows the modification of the cylindrical path, which consists in changing the shape into a sinusoidal shape.

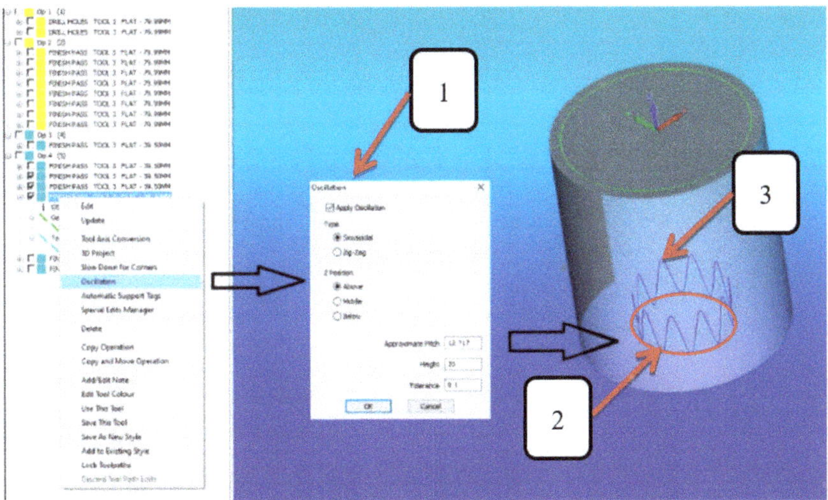

Figure 9. Programming in non-conventional way by adding oscillation frequency and amplitude's high to circular path generated in CAD/CAM system; 1—entering parameters, 2—circular path, 3—the resulting sinusoidal trajectory.

Figure 10 shows how to modify the shape of the machining path (item 2 from Figure 9) to a zig-zag shape path or a sinusoidal path (item 3 from Figure 9).

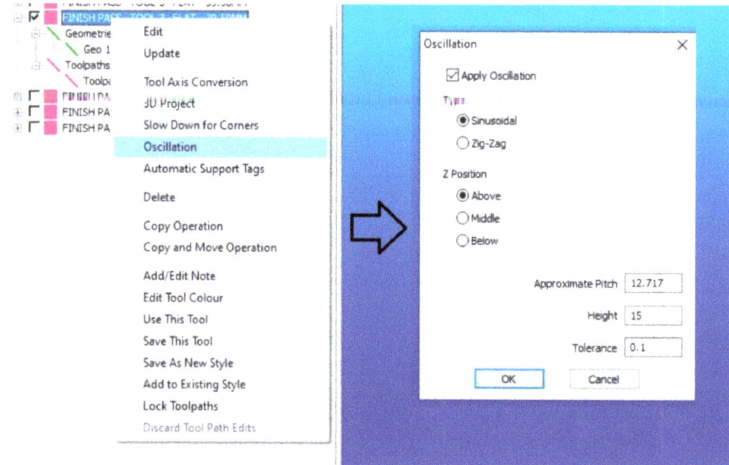

Figure 10. The method of modifying the shape of the machining path in CAD/CAM system Alphacam.

3.4. Setting of Honing Parameters

Automated honing cell should automatically setup the needing abrasive grain trajectories shape, as shown on Figure 9. Construction of honing head should allow for two-way steering of direction of abrasive whetstone and honing head movement (Figure 11) and should have possibilities to receive a machined surface images during process.

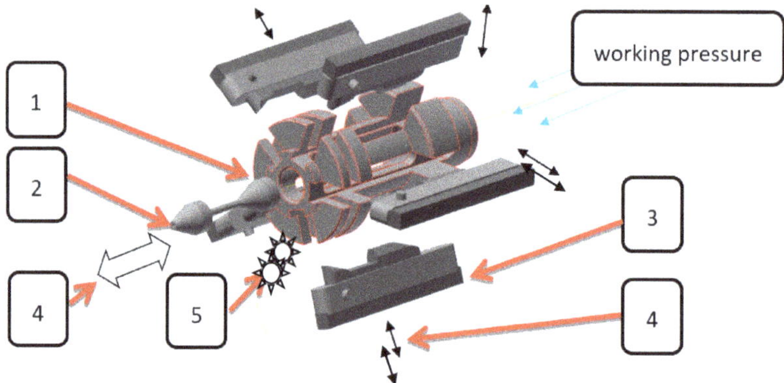

Figure 11. Idea of honing process with abilities to control the abrasive whetstone and honing head movements in both direction: 1—honing head body; 2—expanding pin for abrasive whetstones; 3—abrasive whetstone; 4—possible TWO-WAY direction of movement control; 5—automatic vision system.

An important task is to use appropriate honing parameters that influence the course of the process. Incorrect honing parameters result in an excessive temperature increase and may cause the whetstone wear out in a very short time. Figure 12 shows the whetstone: 1—new, 2—after machining with incorrect pressure at the beginning of honing process, 3—worn out in a very short time.

Figure 12. Machining tool: 1—new whetstone; 2—damaged whetstone; 3—worn whetstone.

It is advantageous to set machining parameters influencing uniform wear of the machining tool.

3.5. Supervision of Honing during Process

A very important issue for the automatic honing cell, in addition to the selection of machining parameters after the initial numerical simulation of the honing process, is to verify, supervise, and correct the obtained machining results in real time of honing process.

3.5.1. Supervising of Surface Textures

Figure 13 shows the surface obtained after honing with a variable value of the linear feed of the honing head. Variable honing kinematics, unlike traditional kinematics, enables the creation of oil scratches with new, and never used earlier, shapes of abrasive grain trajectories. The literature clearly shows the advantages of honing with variable kinematics, which reduces the wear of the machining tool and with surfaces with lower roughness profile parameters, e.g., R_a and R_{pk}.

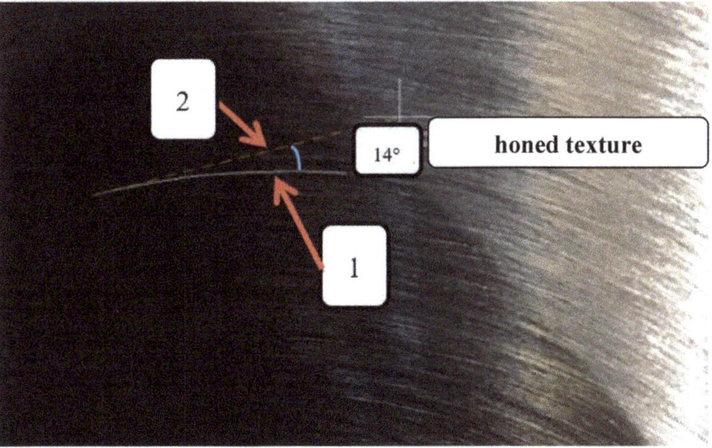

Figure 13. Obtained texture of the honed surface for variable stroke speed of honing head in the range of 1000–3000 mm/min—average value of tangent angle to the grain trajectory of 14°; 1—sample abrasive grain path; 2—tangent line to the abrasive grain path.

Figures 14 and 15 shows examples of surfaces obtained after honing, on which, in addition to measuring the deviation in the shape of roundness and cylindricity and the parameters of the roughness profile, the quality of received texture of machined hole is checked.

Figure 14. Example of a honed surface without texture defects, 1—probe tip.

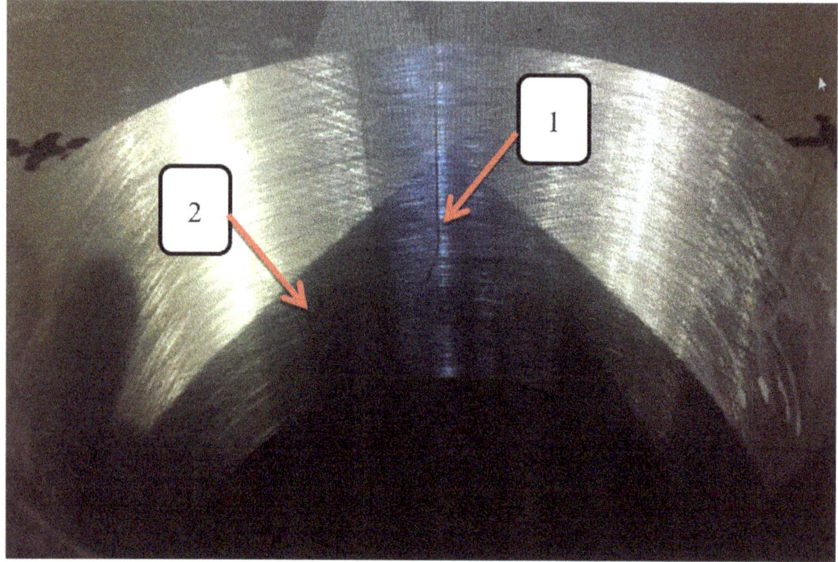

Figure 15. An example of a honed surface with texture defects: 1—scratch; 2—point flaw.

Figure 14 shows the surface after honing, without texture defects.

Figure 15 shows the surface after honing, with texture defects in the form of: 1—scratches, 2—point heterogeneity. The quality of the obtained surface is determined by the homogeneity of the dimensions, shape, and texture of the obtained surface after honing process.

Figure 16 shows the verification of the obtained oil channel pattern in a schematic manner. The verification consists in checking whether the obtained oil channels are continuous, or whether they are clogged with fragments of the honed workpiece's material. The individual layers of neural networks check whether the image fragment shows an oil channel shape is a line, whether it is a curve, whether the shape is broken, or the break is caused by the intersection of oil channels or the presence of the workpiece material in the oil channel. In the event that the verification would confirm a significant number of oil channel breaks, the shape of the grain path should change, and the treatment should ensure the minimum number of places with a break for oil flow.

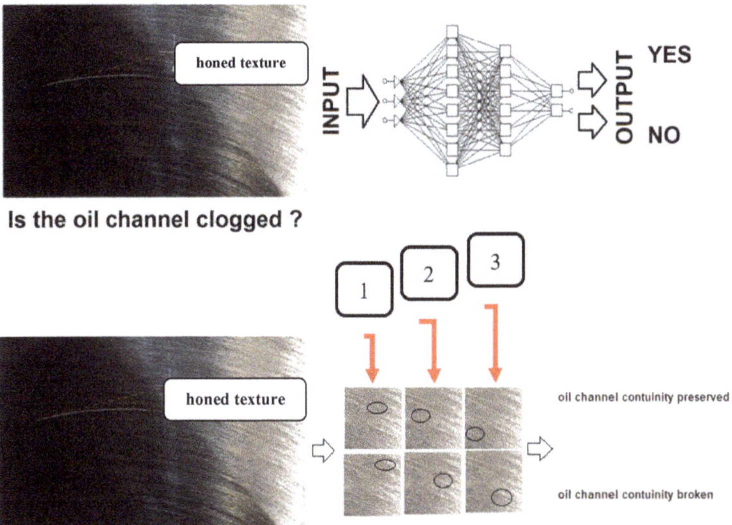

Figure 16. Verification of the shape and it's continuity of oil channels using a neural network. The numbers indicate the stages of the subsequent stages of surface texture verification.

3.5.2. Analysis of Image of Honed Workpieces in Matlab's Image Processing Toolbox

Figure 17 shows a cone representing the HSV color description method. Each color has its own shade, brightness, and value by which the color can be defined.

At the beginning of the honing process, the workpiece temperature is equal to the ambient temperature. During processing, the temperature of the workpiece increases, which causes thermal deformations of the honed hole. The temperature rise of the item can be observed on-line through the Matlab Image Processing Toolbox module.

3.6. Correcting of Honing Parameters

The honing cell should include a CNC honing machine, a thermal infrared camera, a microphone, and modules of e.g., Matlab's software for on-line image and sound level spectrum analysis, an oil cooler and a *CCR* (curve curvature radius) module—the matching module of the shape of the abrasive grain trajectories received on developed surface of the machined workpieces to a certain thickness of the cross section of the honed workpiece.

Figure 18 shows possible actions to be performed during the honing process by automated honing cell: (a) FEM mesh overlay and working pressure setting, (b) determination of thermal conditions, (c) deformation, sound level and image analysis, (d) determination of honing parameters, (e) honing with variable kinematics setting with *CCR* module, (f) receiving variable shape of abrasive grain trajectories optimized to manufacturing conditions.

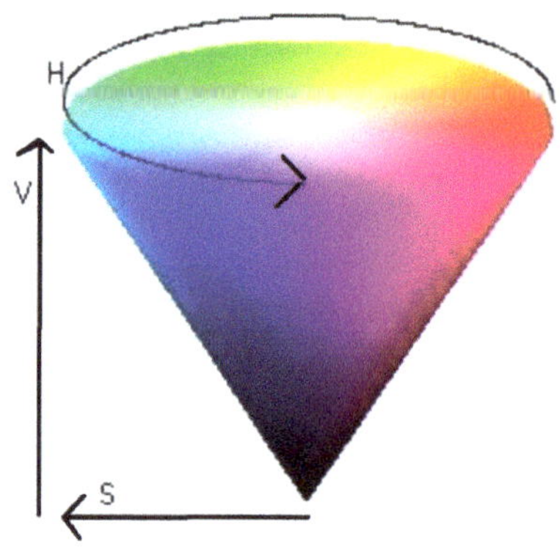

Figure 17. Cone showing the color description method named HSV.

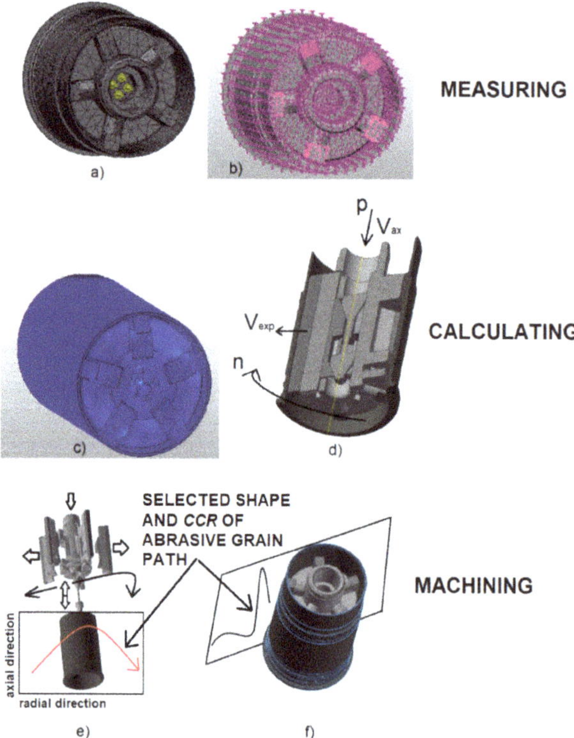

Figure 18. Schematic diagram of honing algorithm: (**a**) FEM mesh overlay and working pressure setting, (**b**) determination of thermal conditions, (**c**) deformation, sound level and image analysis, (**d**) determination of honing parameters, (**e**) honing with variable kinematics setting with *CCR* module, (**f**) receiving variable shape of abrasive grain trajectories optimized to manufacturing conditions.

Figure 19 shows schematically the process parameters verified on-line during the treatment. Depending on the values of the obtained parameters, the values of the machining parameters would be corrected automatically.

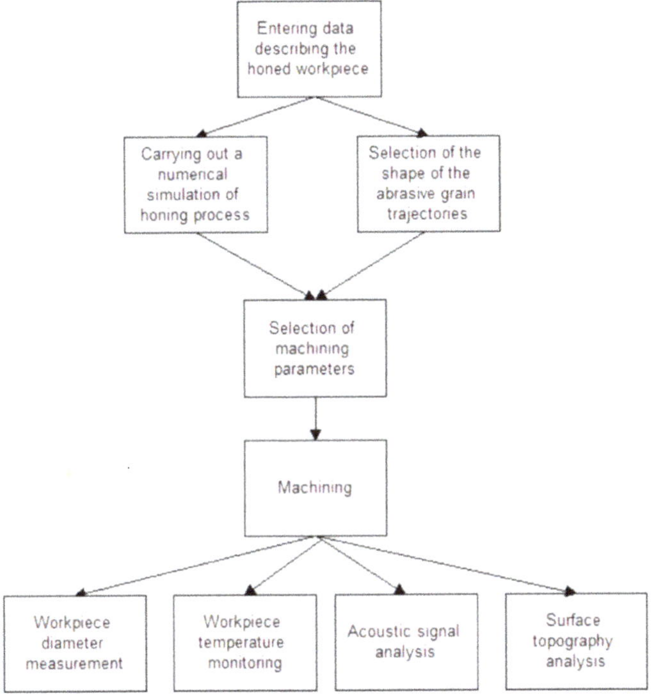

Figure 19. Schematic diagram of automated honing process.

The main factor influencing the differentiation of efficiency of manufacturing is the lack of the need to multiple cool the honed workpieces before the end of the honing process, and lack of the multiple measurements of the obtained diameter and shape deviation of honed hole.

4. Results

Performing numerical simulations makes it easier to plan the machining process. Owing to the simulation, we can find out the size of deformations and stresses occurring during honing, which allows us to decide on the selection of the right machining parameters.

Owing to the sound signal level analysis and thermal image of honed workpieces analysis one can establish the needed parameters of the honing process.

4.1. Stresses and Deformations in Machined Workpieces

In Figure 20 digit 1 indicates the location of the stress measurement, that the honing cell could monitor online. A honing head with the ability to measure the amount of pressure of the whetstone on the treated surface would allow for the verification of the shape deviation of the honed hole.

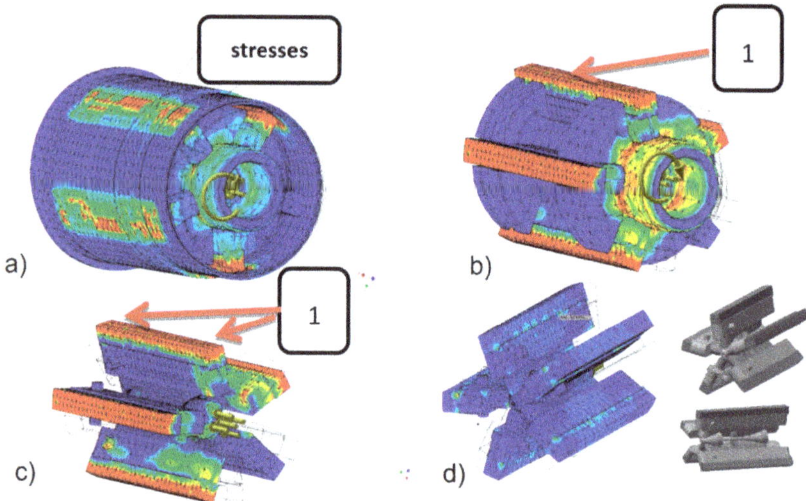

Figure 20. Results of numerical analysis of honing process of cylinder linear with different thickness of cross-section; (**a**) simulation result of the entire assembly, (**b**) assembly simulation result without cylinder linear, (**c**) assembly simulation result without cylinder linear and without honing head body, (**d**) simulation of stresses obtained during honing process.

Figure 21 shows the numerical inhomogeneous simulation values of the deformation of the honed hole obtained during honing of thin-walled workpiece with a variable wall thickness, which shows the actual manufacturing difficulties of this type of workpieces.

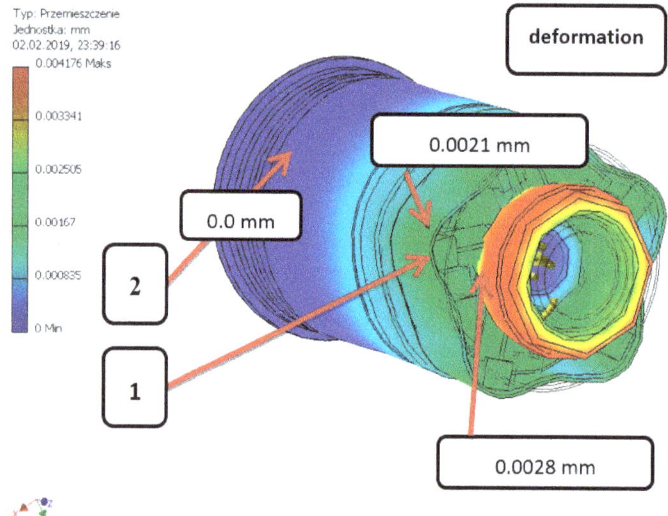

Figure 21. Numerical simulation—deformation of a cylinder with a variable wall thickness: 1—the greatest cylindrical deformation value of honed workpiece; 2—the smallest cylindrical deformation value of honed workpiece.

4.2. Analysis of Sound Signal Level, Received During Honing Process, in Matlab's Audio Toolbox

Figure 22 shows a graph of the intensity of the sound signal, which was obtained during honing with varying kinematics. Measured sound was introduced into the Matlab Audio Toolbox. Distance from peak to peak in horizontal direction, in the diagram shown in Figure 22, indicates the time of the honing cycle (movement of the striking head up and down). Similarly, to Buj-Corral I. [2], it has been verified that the choice of honing parameters is reflected by the amount of sound signal emission (Figure 22), which means that the acoustic signal analysis is a good tool for verifying the honing process. Figure 23 shows the test stand with the equipment for measuring the sound intensity level.

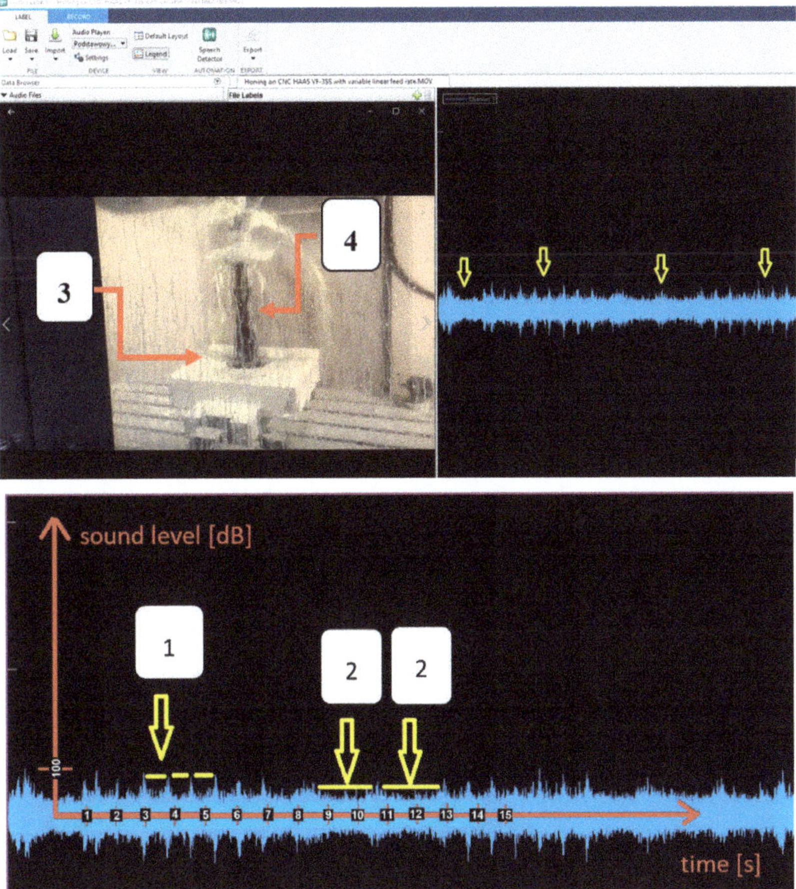

Figure 22. Matlab Audio Labeler—analysis of sound level vs honing process time conducted on CNC vertical milling machine Haas VF-3SS. Honing with variable kinematics condition, with different value of honing head stroke speed; 1—shorter honing cycle time, 2—longer honing cycle time, 3—honed workpiece, 4—honing equipment.

Figure 24 shows the sound pressure level for different mean values of the variable stroke feed of the honing head. Figure 24 shows that the lower value of the sound intensity level is obtained for the mean value range of the variable stroke feed of the honing head.

Figure 23. Test stand: 1 equipment for measuring of the sound intensity level.

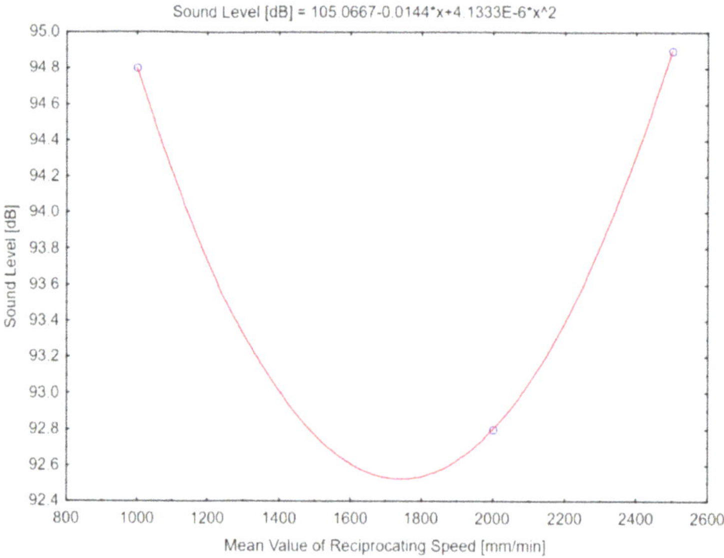

Figure 24. Sound intensity level depending on the average value of the variable head feed, obtained during honing with variable kinematics.

A lower sound level [dB] value means the occurrence of lower cutting forces, associated with the occurrence of lower cutting forces for the middle range of the applied feed.

4.3. Analysis of Image of Honed Workpieces in Matlab's Image Processing Toolbox

Figures 25 and 26 show the temperatures of the processed object obtained during honing. Using the HSV description method, it was proposed to analyze the temperature distribution over the surface of the entire object. Instead of checking the highest temperature to which the honed object has heated up, an analysis of the uniformity of the temperature of the entire object was proposed. As shown in Figure 25, the temperatures analyzed in the Matlab Image Processing Toolbox module (value and degree of heating of the object) are verified using the Hue, Value, and Saturation parameter.

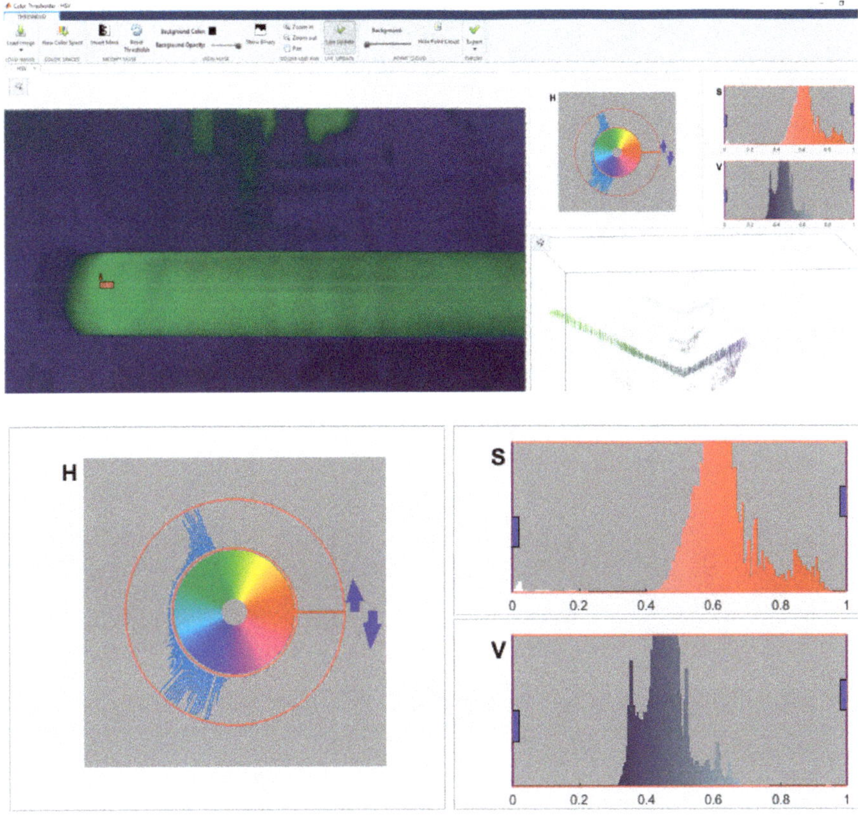

Figure 25. Matlab Image Colour Thresholder—analysis of thermogram of honed workpiece on CNC vertical milling machine Haas VF-3SS (earlier stage of honing (than on Figure 26). H—hue, S—Saturation, and V—Value (HSV).

It can be clearly noticed that in Figure 26, i.e., on the workpiece after processing (heated from honing), the H parameters shift to the right toward the red shade, which means that the workpiece has heated up (the recorded thermogram shows a higher temperature than at the beginning of the treatment). The value and saturation of the color, the S and V parameters, respectively, also shift to the right, which suggests an increase in the temperature value on a larger surface of the object to be polished.

The task of the automatic cell is to monitor the degree of heating of the object, while the H parameter takes the appropriate color value, the S and V parameters respectively will confirm that the object has been heated evenly over the entire surface.

As shown in Figure 26 the hotter the temperature of honed workpiece, the more bright red color. Figure 26 also shows the shifting of the amount of the measured light color to the right, i.e., toward the lighter colors, by means of arrows.

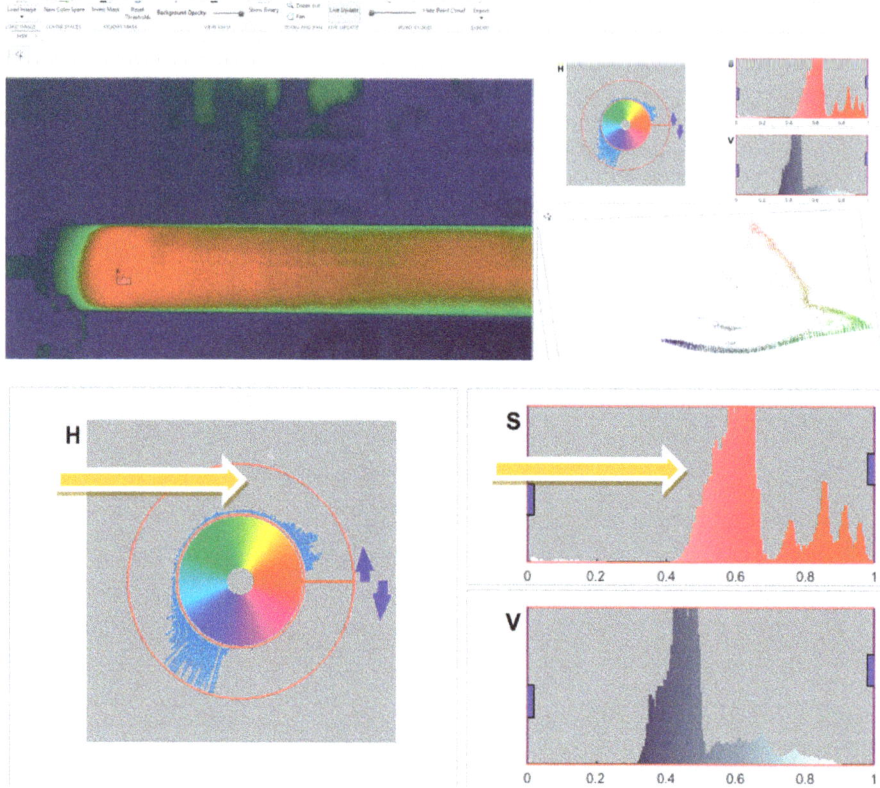

Figure 26. Matlab Image Colour Thresholder—analysis of thermogram of honed part on CNC vertical milling machines Haas VF-3SS (later stage of honing process than shown on Figure 25). H—hue, S—Saturation, and V—Value (HSV).

The analysis of the thermogram, in contrast to the analysis of only the maximum temperature of the workpiece, can provide information about the local temperature increase of the honed workpiece during machining, which provides comprehensive knowledge about the temperature rise in the workpiece for different cross-section thicknesses and for different places on the honed hole.

Figure 27 shows a window view from the FlexSim 2020 simulator in which a simulation of honing on an automated cell was prepared and compared to conventional honing. The automatic cell enables more than 20 times faster production than in the conventional way.

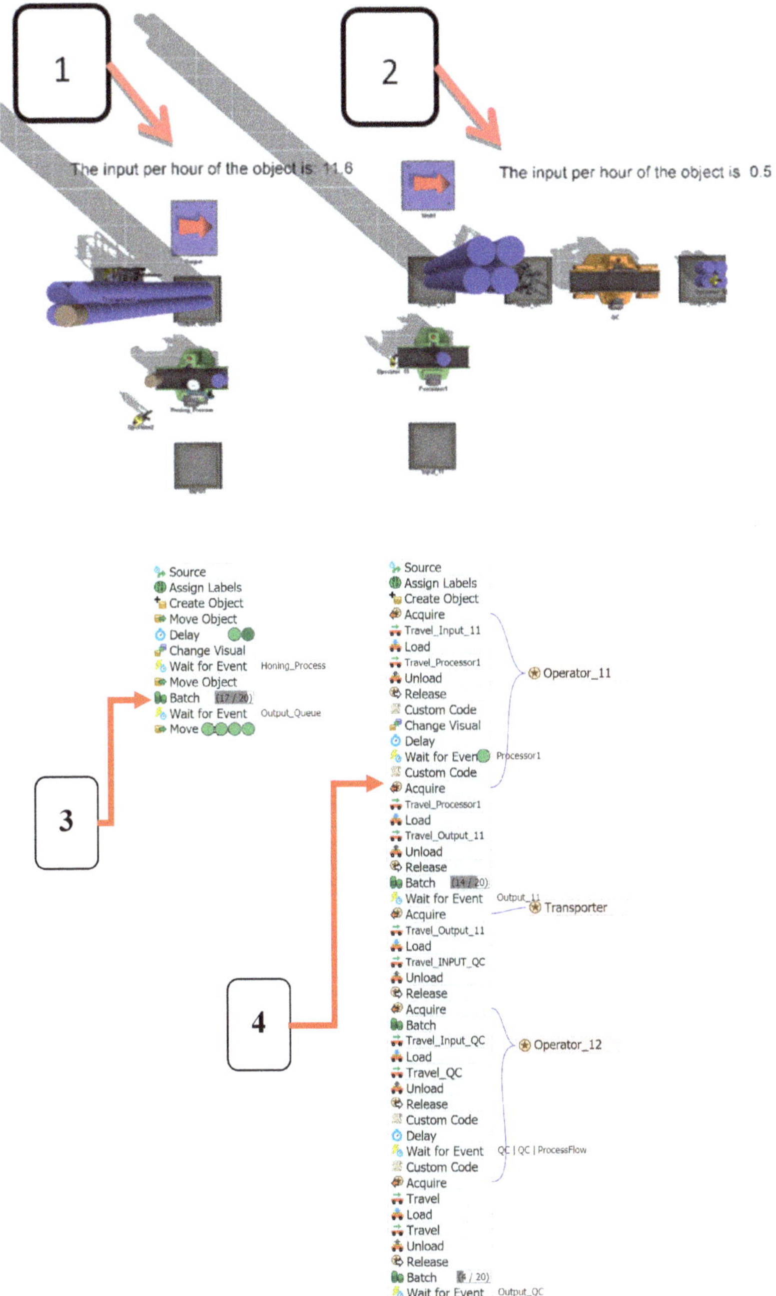

Figure 27. Window from the FlexSim 2020 simulator—comparison of production on a conventional station and on an automatic cell; 1—automated honing cell (efficiency of automated honing cell: 11.6 workpieces/h), 2—conventional honing cell (efficiency of traditional honing: 0.5 workpiece/h). 3 and 4—FlexSim 2020 simulation's algorithms of honing process in Process Flow.

5. Discussion

Building a test stand with all components listed in the article will allow to produce workpieces of complex shapes with different wall thickness and required specifications. The proposed honing cell consists of CNC honing center with honing equipment, thermographic camera, profilographometer, sound intensity measuring device, software for a thermographic camera, software for a roughness gauge, software for a sound level meter, cooling nozzle, measuring instruments—a diaphragm gauge, and computers enabling simultaneous observation of changes of data for individual instruments could enable the automation of honing process and the reduction of manufacturing time.

When changing the machining parameters and when switching the honing process on and off, individual devices had to be turned on, set, turned off, and properly turned on and off for collecting data in a dedicated software. After setting up the tooling, it was necessary to upload the appropriate program to the CNC machine, turn on individual devices and their dedicated software.

A very advantageous solution would be to create a honing cell with all listed elements implemented in honing machine and used simultaneously during honing.

The automatic honing cell could monitor on-line the temperature of the honed workpiece, the pressure of the whetstones obtained during machining, the sound level of honing process, and depending on the honing process conditions, it could automatically, using the *CCR* module, create an abrasive grain path shape with an appropriate radius of curve curvature to improve machining conditions. An important conclusion is the possibility of about a 20-fold increase in the efficiency of serial production of honing thin-walled objects

6. Conclusions

Performing a simulation of machining before the honing process enables the selection of machining parameters before the process begins, it would be particularly valuable to program the shape of the grain trajectory with a specific value of the curve radius adjusted to the honed workpieces.

The measurement of the sound intensity in the honing process is a valuable source of information because the value of the sound level is directly related to the processing conditions, which is easily noticeable in the way that the lower sound level corresponds to the lower cutting forces occurring at a given processing time.

The temperature increase of the honed workpieces can be monitored during the honing process, the advantage of the HSV method is the fact that we analyze the temperature distribution over the entire smoothed surface, and not only collect information about the maximum temperature obtained, this method has many advantages, mainly that it can control the temperature distribution in such a way that the honing process heats the workpiece uniformly.

Supervision and control of the honing process enables about twenty-fold reduction of the time needed to perform honing machining of thin-walled objects and of variable thicknesses of workpieces, because of the elimination of the need to divide the processing into several stages, followed by cooling to the ambient temperature, in which diameter of the honed hole can be measured.

The analysis of the surface texture obtained during honing process, with the use of the neural networks, allows for quick verification whether the obtained surface has a uniform texture shape of the obtained oil channels or whether additional corrective machining passes of honing head are required.

Author Contributions: Conceptualization, P.S.; methodology, P.S.; software, P.S.; validation, A.B.; formal analysis, A.B.; investigation, P.S.; resources, P.S.; data curation, P.S.; writing—original draft preparation, P.S.; writing—review and editing, A.B.; visualization, P.S.; supervision, A.B.; project administration, A.B. All authors have read and agreed to the published version of the manuscript.

Funding: This research received no external funding.

Acknowledgments: The authors thank for co-financing the Mazovia/0127/19 project by the National Center for Research and Development.

Conflicts of Interest: The authors declare no conflict of interest. The funders had no role in the design of the study; in the collection, analyses, or interpretation of data; in the writing of the manuscript, or in the decision to publish the results.

References

1. Barylski, A.; Sender, P. Research on the increase in diameter and temperature of objects during honing long holes in production conditions. *Mechanic* **2014**, *9*, 34–43.
2. Buj, I.; Álvarez-Flórez, J.; Domínguez-Fernández, A. Acoustic emission analysis for the detection of appropriate cutting operations in honing processes. *Mech. Syst. Signal Process.* **2018**, *99*, 873–885. [CrossRef]
3. Buj-Corral, I.; Vivancos-Calvet, J.; Setien, I.; Sebastian, M.S. Residual stresses induced by honing processes on hardened steel cylinders. *Int. J. Adv. Manuf. Technol.* **2016**, *88*, 2321–2329. [CrossRef]
4. González-Rojas, H.A.; Vivancos-Calvet, J.; Salcedo, M.C. Thermal Analysis of Honing Process. *Mater. Sci. Forum* **2006**, *526*, 235–240. [CrossRef]
5. Sender, P.G. Curve Curvature Analysis of a Grain Trajectories in Variable Honing of Cylindrical Holes of Thin Wall Cylinder Liners as a Honing Process Optimization Strategy. In *Innovations Induced by Research in Technical Systems*; Majewski, M., Kacalak, W., Eds.; Springer: Cham, Switzerland, 2020; pp. 81–93.
6. Sender, P. Influence of the abrasive grain trajectory on the machining of thin-walled cylinder liners of internal combustion engines with variable kinematics of honing. *Autobus Tech. Eksploat. Syst. Transp.* **2018**, *19*, 634–641. [CrossRef]
7. Sender, P. Variable Kinematics of Honing Process—Influence on Machined Workpiece. In Proceedings of the Trends in the Development of Machinery and Associated Technology, TMT 2018, Karlovy Vary, Czech Republic, 18–22 September 2018. Available online: https://www.tmt.unze.ba/proceedings2018.php (accessed on 9 September 2018).
8. Sender, P.G. Numerical simulations of honing process of thin wall cylinder liners, with constant and with variable thickness of wall of honed parts. *Mach. Technol. Mater.* **2019**, *13*, 338–344.
9. Yokoyama, K.; Ichimiya, R. *Analysis of Thermal Deformation of Workpiece in Honing Process: 3rd Report Numerical Analyses of Cylindrical and Non-Cylindrical Workpieces*; Bulletin of the Japan Society of Precision Engineering: Tokyo, Japan, 1982.
10. Yokoyama, K.; Ichimiya, R.; Iwata, K.; Moriwaki, T. *Analysis of Thermal Deformation on Workpiece on Honing Process: 5th Report Thermal Effects Due to Heat Capacity of Workpiece and Kind of Honing Stone*; Bulletin of the Japan Society of Precision Engineering: Tokyo, Japan, 1985.
11. Jatti, P.; Mench, R.G. Developing an auto sizing system for vertical honing machine. *Int. J. Recent Res. Civ. Mech. Eng.* **2015**, *1*, 6–15.
12. Deja, M.; List, M.; Lichtschlag, L.D.; Uhlmann, E. Thermal and technological aspects of double face grinding of Al2O3 ceramic materials. *Ceram. Int.* **2019**, *45*, 19489–19495. [CrossRef]
13. Yuan, S.; Huang, W.; Wang, X. Orientation effects of micro-grooves on sliding surfaces. *Tribol. Int.* **2011**, *44*, 1047–1054. [CrossRef]
14. Khanov, A.M.; Muratov, K.R.; Gaszew, E.A.; Pepelyszewie, A.V. Kinematics of honing methods. *Russ. Acad. Sci.* **2011**, *13*, 4.
15. Podgaetski, M.; Scherbina, K. Kinematics of cutting holes in honing spiral spring hone. *Autobus Tech. Eksploat. Syst. Transp.* **2015**, *6*, 2409–9392.
16. Podgaetski, M.; Sczerbina, К.К. УТВОРЕННЯСКЛАДНОЇ ТРАЄКТОРІЇ РУХУ ЗЕРНА ПРИ ХОНІНГУВАННІ ОТВОРІВ. Кіровоградський національний технічний університет, м. Кіровоград, Україна, n° УДК621.923.5. Available online: https://core.ac.uk/download/pdf/84824964.pdf (accessed on 27 October 2020).
17. Yousfi, M.; Mezghani, S.; Demirci, I.; El Mansori, M. Tribological performances of elliptic and circular texture patterns produced by innovative honing process. *Tribol. Int.* **2016**, *100*, 255–262. [CrossRef]
18. Ogorodov, V.A. Hole shaping in the honing of thin-walled cylinders. *Russ. Eng. Res.* **2017**, *37*, 549–553. [CrossRef]
19. Schmitt, R.; König, N.; Zheng, H. Machine integrated optical measurement of honed surfaces in presence of cooling lubricant. *J. Phys. Conf. Ser.* **2011**, *311*, 012007. [CrossRef]

20. Tripathi, B.N.; Singh, N.K.; Vates, U.K. Surface Roughness Influencing Process Parameters & Modeling Techniques for Four Stroke Motor Bike Cylinder Liners during Honing: Review. *Int. J. Mech. Mech. Eng.* **2015**, *15*, 1.

21. Voronov, S.A. *Development of Mathematical Methods for the Analysis of Dynamics of Hole Finishing Processes*; Moscow State Technical University: Moscow, Russia, 2008.

22. Wang, J.; Shao, Y.; Zhu, X. Kinematics analysis and experimental study on ultrasonic vibration honing. In Proceedings of the International Technology and Innovation Conference 2009 (ITIC 2009), Xi'an, China, 12–14 October 2009. [CrossRef]

23. Xi, C.; Hu, X.; Zhang, Z. Research for cylindricity prediction model of inner-hole honing. In Proceedings of the 2011 Second International Conference on Mechanic Automation and Control Engineering (MACE), Inner Mongolia, China, 15–17 July 2011; pp. 1506–1509.

24. Zhang, Y.; Yang, Y.; Niu, J.; Gong, J. (Eds.) Study on the Impact of Honing Machine Reciprocating Reversing Acceleration upon Reticulate Pattern Trajectory. In Proceedings of the 1st International Conference on Mechanical Engineering and Material Science, Shanghai, China, 18–20 December 2020; pp. 17–20.

25. Arantes, L.J.; Fernandes, K.A.; Schramm, C.R.; Leal, J.E.S.; Piratelli-Filho, A.; Franco, S.D.; Arencibia, R.V. The roughness characterization in cylinders obtained by conventional and flexible honing processes. *Int. J. Adv. Manuf. Technol.* **2017**, *93*, 635–649. [CrossRef]

26. Buj-Corral, I.; Vivancos-Calvet, J.; Rodero-De-Lamo, L.; Marco-Almagro, L. Comparison between Mathematical Models for Roughness Obtained in Test Machine and in Industrial Machine in Semifinish Honing Processes. *Procedia Eng.* **2015**, *132*, 545–552. [CrossRef]

27. Deepak Lawrence, K.; Ramamoorthy, B. Multi-surface topography targeted plateau honing for the processing of cylinder liner surfaces of automotive engines. *Appl. Surf. Sci.* **2016**, *365*, 19–30. [CrossRef]

28. Kadyrov, R.; Charikov, P.N.; Pryanichnikova, V.V. Honing process optimization algorithms. *IOP Conf. Series: Mater. Sci. Eng.* **2018**, *327*, 022052. [CrossRef]

29. Pawlus, P.; Michalski, J.; Reizer, R. Progress in cylinder honing. In *Part I: Honing of Blind Holes*; Rusek, P., Ed.; Instytut Zaawansowanych Technologii Wytwarzania: Kraków, Poland, 2012.

30. Schmitt, C.; Klein, S.; Bähre, D. An Introduction to the Vibration Analysis for the Precision Honing of Bores. *Procedia Manuf.* **2015**, *1*, 637–643. [CrossRef]

31. Gouskov, A.; Voronov, S.A.; Butcher, E.A.; Sinha, S. *Non-Conservative Oscillations of a Tool for Deep Hole Honing*; Elsevier: Amsterdam, The Netherlands, 2006; pp. 685–708.

32. Iskra, A.; Kałużny, J. Influence of the actual shape of the piston side surface on the parameters of the oil film. *J. KONES Intern. Combust. Engines* **2000**, *7*, 1–2.

33. Schmitt, C.; Bähre, D. Analysis of the Process Dynamics for the Precision Honing of Bores. *Procedia CIRP* **2014**, *17*, 692–697. [CrossRef]

34. Schmitt, C.; Bähre, D. An Approach to the Calculation of Process Forces During the Precision Honing of Small Bores. *Procedia CIRP* **2013**, *7*, 282–287. [CrossRef]

35. Zhang, X.; Wang, X.; Wang, D.; Yao, Z.; Xi, L.; Wang, X. Methodology to Improve the Cylindricity of Engine Cylinder Bore by Honing. *J. Manuf. Sci. Eng.* **2016**, *139*, 031008. [CrossRef]

36. Grzesik, W. Influence of surface textures produced by finishing operations on their functional properties. *J. Mach. Eng.* **2016**, *16*, 15–23. [CrossRef]

37. Babiczew, A.P.; Poljanchikov, J.I.; Slavin, A.V. *Gładzenie*. Министерство Образования и Науки Российскойфедерации. Волгогр. Гос. Архит.-Строит. ун-т; Донской гос. техн. ун-т.: Волгоград, Russia, 2013.

38. Chavan, P.S.; Harne, M.S. Effect of Honing Process Parameters on Surface Quality of Engine Cylinder Liners. *Int. J. Eng. Res. Technol.* **2013**, *2*, 4.

39. Gashev, E.A.; Muratov, K.R. Tool motion in the honing of cylindrical surfaces. *Russ. Eng. Res.* **2014**, *34*, 268–271. [CrossRef]

40. Goeldel, B.; El Mansori, M.; Dumur, D. Macroscopic simulation of the liner honing process. *CIRP Ann.* **2012**, *61*, 319–322. [CrossRef]

41. Goeldel, B.; El Mansori, M.; Dumur, D. Simulation of Roughness and Surface Texture Evolution at Macroscopic Scale During Cylinder Honing Process. *Procedia CIRP* **2013**, *8*, 27–32. [CrossRef]

42. Grabon, W.; Pawlus, P.; Wos, S.; Koszela, W.; Wieczorowski, M. Effects of honed cylinder liner surface texture on tribological properties of piston ring-liner assembly in short time tests. *Tribol. Int.* **2017**, *113*, 137–148. [CrossRef]

43. Jocsak, J. The Effects of Surface Finish on Piston Ring-Pack Performance in Advanced Reciprocating Engine Systems. Master's Thesis, Massachusetts Institute of Technology, Boston, MA, USA, 2005.

44. Johansson, S.; Nilsson, P.H.; Ohlsson, R.; Anderberg, C.; RosénB, -G. Optimization of the Cylinder Liner Surface for Reduction of Oil Consumption. In *World Tribology Congress III*; Imperial College Press: London, UK, 2005; pp. 559–560.

45. Johansson, S.; Nilsson, P.H.; Ohlsson, R.; Anderberg, C.; Rosén, B.-G. New cylinder liner surfaces for low oil consumption. *Tribol. Int.* **2008**, *41*, 854–859. [CrossRef]

46. Khanov, A.M.; Gashev, E.A.; Muratov, K.R. Formation of raster trajectories in the honing of cylindrical surfaces. *Russ. Eng. Res.* **2013**, *33*, 423–426. [CrossRef]

47. Khanov, A.M.; Muratov, K.R.; Gashev, E.A.; Muratov, R.A. Kinematic potential of honing machines. *Russ. Eng. Res.* **2011**, *31*, 607–609. [CrossRef]

48. Knoll, G.; Rienacker, A. *Tribology in Automotive Engine Aplications*; Technical report for Institute for Machine Elements and Design; University of Kassel: Kassel, Germany, 2011.

49. Mezghani, S.; Demirci, I.; Yousfi, M.; El Mansori, M. Mutual influence of crosshatch angle and superficial roughness of honed surfaces on friction in ring-pack tribo-system. *Tribol. Int.* **2013**, *66*, 54–59. [CrossRef]

50. Muratov, K.R.; Gashev, E.A. Methods of precision hole honing. Ph.D. Thesis, Perm National Research Polytechnic University, Perm Krai, Russia, 2014.

51. Obara, R.; Souza, R.M.; Tomanik, E. Quantification of folded metal in cylinder bores through surface relocation. *Wear* **2017**, *384*, 142–150. [CrossRef]

52. Pawlus, P.; Cieslak, T.; Mathia, T. The study of cylinder liner plateau honing process. *J. Mater. Process. Technol.* **2009**, *209*, 6078–6086. [CrossRef]

53. Buj, I.; Vivancos-Calvet, J. Roughness variability in the honing process of steel cylinders with CBN metal bonded tools. *Precis. Eng.* **2011**, *35*, 289–293. [CrossRef]

54. Buj-Corral, I.; Sivatte-Adroer, M.; Llanas-Parra, X. Adaptive indirect neural network model for roughness in honing processes. *Tribol. Int.* **2020**, *141*, 105891. [CrossRef]

55. Goeldel, B.; Voisin, J.; Dumur, D.; El Mansori, M.; Frabolot, M. Flexible right sized honing technology for fast engine finishing. *CIRP Ann.* **2013**, *62*, 327–330. [CrossRef]

56. Buyukli, I.; Kolesnik, V. Improving accuracy of holes honing. *Odes'kyi Polite-Universytet Pr.* **2015**, *215*, 34–43. [CrossRef]

57. Bouassida, H. Lubricated piston ring cylinder liner contact: Influence of the Liner Microgeometry. INSA de Lyon, Dostępne na Stronie Internetowej. 2014. Available online: http://theses.insa-lyon.fr/publication/2014ISAL0088/these.pdf (accessed on 1 January 2019).

58. Corral, I.B.; Vivancos-Calvet, J.; Salcedo, M.C. Use of roughness probability parameters to quantify the material removed in plateau-honing. *Int. J. Mach. Tools Manuf.* **2010**, *50*, 621–629. [CrossRef]

59. Entezami, S.S.; Farahnakian, M.; Akbari, A. Experimental Study of Effective Parameters on Honing Process of Cast Iron Cylinder. *J. Mod. Process. Manuf. Prod.* **2015**, *4*, 3.

60. Karpuschewski, B.; Welzel, F.; Risse, K.; Schorgel, M. Reduction of Friction in the Cylinder Running Surface of Internal Combustion Engines by the Finishing Process. *Procedia CIRP* **2016**, *45*, 87–90. [CrossRef]

61. Mansori, M.E.; Goeldel, B.; Sabri, L. Performance impact of honing dynamics on surface finish of precoated cylinder bores. *Surf. Coatings Technol.* **2013**, *215*, 334–339. [CrossRef]

62. Brush Research Manufacturing Co. Inc. A Study of a Cylinder Wall Micro-Structure. 03/18/2018. Available online: http://info.brushresearch.com/study-of-cylinder-wall-micro-structure (accessed on 1 January 2019).

63. Dahlmann, D.; Denkena, B. Hybrid tool for high performance structuring and honing of cylinder liners. *CIRP Ann.* **2017**, *66*, 113–116. [CrossRef]

64. Demirci, I.; Mezghani, S.; Yousfi, M.; El Mansori, M. Impact of superficial surface texture anisotropy in helical slide and plateau honing on ring-pack performance. *Proc. Inst. Mech. Eng. Part J J. Eng. Tribol.* **2016**, *230*, 1030–1037. [CrossRef]

65. Deshpande, A.K.; Bhole, H.A.; Choudhari, L.A. Analysis of Super-Finishing Honing Operation with Old and New Plateau Honing Machine Concept. *Int. J. Eng. Res. Gen. Sci.* **2015**, *3*, 812–818.

66. Fiat Chrysler America. *Chrysler IIIH Engine Assembly Manual DRAFT*; Fiat Chrysler America: Auburn Hills, MI, USA, 2016.

67. Grabon, W.; Pawlus, P.; Sep, J. Tribological characteristics of one-process and two-process cylinder liner honed surfaces under reciprocating sliding conditions. *Tribol. Int.* **2010**, *43*, 1882–1892. [CrossRef]

68. Michalski, J.; Woś, P. The effect of cylinder liner surface topography on abrasive wear of piston–cylinder assembly in combustion engine. *Wear* **2011**, *271*, 582–589. [CrossRef]

69. Reizer, R.; Pawlus, P.; Galda, L.; Grabon, W.; Dzierwa, A. Modeling of worn surface topography formed in a low wear process. *Wear* **2012**, *278*, 94–100. [CrossRef]

70. Kim, J.-K.; Xavier, F.-A.; Kim, D.-E. Tribological properties of twin wire arc spray coated aluminum cylinder liner. *Mater. Des.* **2015**, *84*, 231–237. [CrossRef]

71. Kapoor, J. Parametric Investigations into Bore Honing through Response Surface Methodology. *Mater. Sci. Forum* **2014**, *808*, 11–18. [CrossRef]

72. KS Motor Service International GmbH. Honing of Gray Cast Iron Cylinder Blocks. Available online: http://file.seekpart.com/keywordpdf/2011/3/30/2011330113527501.pdf (accessed on 18 March 2018).

73. Mezghani, S.; Demirci, I.; Zahouani, H.; El Mansori, M. The effect of groove texture patterns on piston-ring pack friction. *Precis. Eng.* **2012**, *36*, 210–217. [CrossRef]

74. Reizer, R.; Pawlus, P. 3D surface topography of cylinder liner forecasting during plateau honing process. *J. Phys. Conf. Ser.* **2011**, *311*, 1–6. [CrossRef]

75. Pimpalgaonkar, M.H.; Laxmanrao, G.R.; Laxmanrao, A.D.E.S. A review of optimization process parameters on honing machine. *Int. J. Mech. Prod. Eng.* **2013**, *1*, 5.

76. Qin, P.-P.; Yang, C.-L.; Huang, W.; Xu, G.-W.; Liu, C.-J. Honing Process of Hydraulic Cylinder Bore for Remanufacturing. In Proceedings of the 4th International Conference on Sensors, Measurement and Intelligent Materials, Shenzhen, China, 27–28 December 2016; p. 2352.

77. Sabri, L.; Mezghani, S.; El Mansori, M.; Le Lan, J.-V.; Negro, T.D. 3D Multi-Scale Topography Analysis in Specifying Quality of Honed Surfaces. In Proceedings of the 9th Biennial Conference on Engineering Systems, Haifa, Israel, 7–9 July 2008; pp. 225–232.

78. Polyanchikov, Y.N.; Plotnikov, A.L.; Polyanchikova, M.Y.; Kursin, O.A.; Leshukov, A.V. Honing with increase in the cutting speed. *Russ. Eng. Res.* **2008**, *28*, 727–728. [CrossRef]

79. Yousfi, M. Tribofunctional Study of Low-Friction Engine Liner Textures Generated by Honing Process. 2014. Available online: https://pastel.archives-ouvertes.fr/tel-01148194 (accessed on 5 January 2019).

80. Yousfi, M.; Mezghani, S.; Demirici, I.; Mansori, E.M. (Eds.) Comparative study between 2D and 3D characterization methods for cylinder liner plateau honed surfaces. In Proceedings of the 42nd North American Manufacturing Research Conference, Detroit, MI, USA, 9–13 June 2014; pp. 1–7.

81. Yousfi, M.; Mezghani, S.; Demirci, I.; Mansori, E.M.; Zahouani, H. Energy efficiency optimization of engine by frictional reduction of functional surfaces of cylinder ring-pack system. *Tribol. Int.* **2013**, *59*, 240–247.

82. Yousfi, M.; Mezghani, S.; Demirci, I.; Mansori, E.M. Generation of Circular and Elliptic Low-Friction Texture Patterns by Honing Process. Available online: https://www.researchgate.net/profile/Mohammed_Yousfi4/publication/305045271_Generation_of_circular_and_elliptic_lowfriction_texture_patterns_by_honing_process/links/577ff87d08ae5f367d370b77/Generation-of-circular-and-elliptic-low-friction-texture-patterns-by-honing-process.pdf (accessed on 7 September 2015).

83. Yousfi, M.; Mezghani, S.; Demirci, I.; El Mansori, M. Mutual Effect of Groove Size and Anisotropy of Cylinder Liner Honed Textures on Engine Performances. *Adv. Mater. Res.* **2014**, *966*, 175–183. [CrossRef]

84. Yousfi, M.; Mezghani, S.; Demirci, I.; El Mansori, M. Smoothness and plateauness contributions to the running-in friction and wear of stratified helical slide and plateau honed cylinder liners. *Wear* **2015**, *332*, 1238–1247. [CrossRef]

85. Yousfi, M.; Mezghani, S.; Demirci, I.; Mansori, E.M. Texturation Mécanique Antifriction Par Rodage Du tribo-système Segment-Cylindre. Présentation du Projet D'acquisition d'un tribo-simulateur du Fonctionnement Moteur. In Proceedings of the 28 Journees Internationales Francophones de Tribologie, Lyon, France, 27–29 April 2016.

86. Ozdemir, M.; Korkmaz, M.E.; Gunay, M.B. Optimization of surface roughness in honing of engine cylinder liners witch SiC honing stones. In Proceedings of the 1st International Conference on Engeneering Technology and Applied Sciences, Afyonkarahisar, Turkey, 21–22 April 2016.

87. Buj-Corral, I.; Vivancos-Calvet, J.; Coba-Salcedo, M. Modelling of surface finish and material removal rate in rough honing. *Precis. Eng.* **2014**, *38*, 100–108. [CrossRef]
88. Wang, Q.; Feng, Q.; Li, Q.; Zu Ren, C. The Experimental Investigation of Stone Wear in Honing. *Key Eng. Mater.* **2011**, *487*, 462–467. [CrossRef]
89. Voronov, S.A.; Gouskov, A.; Bobrenkov, O.A. Modelling of bore honing. *Int. J. Mechatronics Manuf. Syst.* **2009**, *2*, 566. [CrossRef]
90. Zhang, X.; Zhu, X.; Cheng, L. The Influence Study of Ultrasonic honing parameters to workpiece surface temperature. *MATEC Web Conf.* **2016**, *45*, 4008. [CrossRef]

Publisher's Note: MDPI stays neutral with regard to jurisdictional claims in published maps and institutional affiliations.

 *machines* 

Article

Analysis of the Possibility of Using Wavelet Transform to Assess the Condition of the Surface Layer of Elements with Flat-Top Structures

Paweł Karolczak, Maciej Kowalski and Magdalena Wiśniewska *

Department of Mechanical Engineering, Wrocław Uniwersity of Science and Technology, Ignacego Łukasiewicza 5 Street, 50-370 Wrocław, Poland; pawel.karolczak@pwr.edu.pl (P.K.); maciej.kowalski@pwr.edu.pl (M.K.)
* Correspondence: m.wisniewska@pwr.edu.pl

Received: 8 September 2020; Accepted: 20 October 2020; Published: 22 October 2020

Abstract: The paper focused on a topic related to the possibilities of using wavelet analysis to evaluate the changes in the geometrical structures of the surfaces arising during the honing process with whetstones with variable granularity. The cylinder liners of the combustion engine are machined elements. The basics of the wavelet analysis and the differences between filtering with standardized filters (e.g., Gauss filter), Fourier analysis, and the analysis of the results obtained when measuring the surface roughness with other wavelets were described. Trials of honing four cylinder liners were carried out. Roughness measurements of 3D spatial structures of the prepared liners were made. The principle of selecting wavelets for roughness assessment of structures with cross-hatch pattern was described. Roughness structures generated on the honed surfaces of cylinder liners were assessed using Gaussian filtration and Morlet, Daubechies Db6, and Mexican hat wavelets. In order to demonstrate the differences generated when the Gaussian filtration and selected wavelets were used on surface structures, the surfaces obtained with the use of these filtering tools were subtracted from each other, which allowed obtaining information about the changes occurring on the assessed surfaces, which were generated after the use of various filtering tools. For the assessed surfaces, during the subtraction operation, the mean square error was calculated, informing about the degree of similarity of both compared surfaces. The result of the work carried out is the creation of basic recommendations for the selection of wavelets when assessing honed surfaces with different degrees of regularity of the traces generated on them.

Keywords: wavelet analysis; decomposition of signals; honing; surface roughness

---

1. Introduction

Nowadays, machining remains the basic manufacturing technique in the industry. The possibilities it offers, in terms of versatility and availability, place machining in the first place, among other techniques. The rapid development of automation and digitalization in machining will ensure its leading position in the near and, probably, in the distant future. Current trends in the industry require maximum flexibility of the production from the manufacturers. Development of Industry 4.0 and many other factors proves that machining is and will be the main choice among other manufacturing techniques.

Wherever production appears, there is also a need to control its effects. Roughness, next to the waviness, is one of the most basic features of the surfaces obtained in the machining process. Roughness measurement allows us to assess the quality of the surface of a given object and classify it as a product needing further machining or as a finished product.

For most applications, linear measurement is perfectly sufficient. This refers especially to one-directional structures, whose surface topography is rather repetitive. However, in some cases, it becomes necessary to use 3D roughness measurement. This enables not only a better understanding of the nature of a surface itself, but it is generally needed when phenomena of contact between two elements are involved, and we cannot limit ourselves only to the analysis of a single profile [1].

Measurement of the surface geometric structure in 3D provides us much important information about it. Modern measurement systems provide the user with enormous possibilities of analysis and facilitate it, for instance, through the implementation of tools for graphical presentation of the results. The obtained results are usually affected by errors resulting from the method itself or the measurement conditions [2]. Many kinds of filters are used to screen out irregularities, of which Gaussian filter is most common. A standardized Gaussian filter is widely described in the literature, in terms of interpretation and calculations relatively easy in use, and most likely used by all devices for the analysis of the geometrical structure. It is calculated based on the Fourier transform, which is the basic tool for analysis and signal processing. The Fourier transform itself is a great tool in the case of stationary signals, and a Gaussian filter is a great tool for general filtering. However, in the case of non-stationary signals or when we want to filter specific parameters or analyze the results for specific properties, they lose their relevance. This is where the wavelet transform comes in handy. It is a transformation similar to Fourier's transformation, but with the main difference being the so-called "kernel transformation". When the Fourier transform is based on the kernel of the sinusoidal function, the wavelet transform uses wavelets, which are is an infinite number. This gives an enormous number of possibilities for signal analysis.

2. Wavelet Analysis

In order to analyze the results in a broad sense, various tools and methods are used to help the user obtain the desired information. The obtained measurement results are often burdened with measurement errors or disturbances, making it difficult to obtain a real picture of the situation. For this purpose, various types of filtration are used to separate the above-mentioned disturbance. In the case of signal analysis of devices used for roughness measurement, methods, such as Gaussian filtration or wavelet analysis, are used and have been briefly described in this chapter.

Both the Gaussian filter and the wavelet analysis are based on Fourier transform, although the Gaussian filter is being calculated on its basis; meanwhile, wavelet transform is, in a way, its extension. It is a transformation similar to the Fourier transform, relying just like it on the scalar product operation.

The Fourier transform is a linear operator defined on certain functional spaces. The elements of these spaces are functions of n real number. In Fourier transform analysis, the harmonic sine wave and cosine harmonic wave are multiplied by a signal. The final integration provides guidance on the signal for a given frequency. A classic example of this is the signal spectrum, a signal's energy in a given point in its frequency domain [3].

By changing the time-value system for the frequency-value system, valuable information about the time is lost during the transformation when the event occurs (Figure 1).

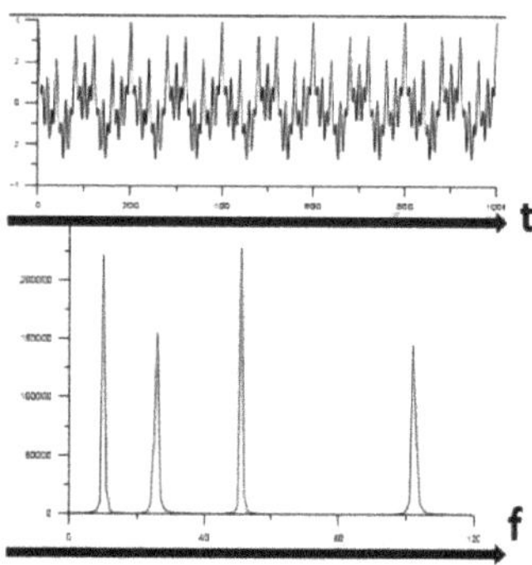

Figure 1. Exemplary signal and its transformation using the Fourier transform [4].

The analysis of the transformed signal shows that the tested signal consists of four harmonic frequencies; however, it cannot be deducted from it how each of the components changes in time.

The Fourier transform is reversible, so by having the same *F(u)* transform, it is possible to determine the original signal before transformation [3].

Intensive development of the methodology related to wavelet transform occurred at the beginning of the 20th century. Wavelet transform found broad applications across many branches of science, mainly due to the possibilities it offers. The wavelet transform, as previously mentioned, is similar to the Fourier transform, as it is based on the use of the scalar product of a given signal and a part called the transformation kernel. The main difference between these two is precisely the kernel. In the Fourier transform, the sinusoidal function is used as a kernel, which means that every function will be represented continuously by one selected frequency. Whereas, in the wavelet transform, the kernel is a function called wavelet, which fulfills the requirements of the time-frequency analysis [5].

Unlike the wave, wavelet has continuous oscillatory waveforms, which have different durations and spectrum. It is finite energy that is concentrated around one point and has a mean value of zero. There is an infinite number of wavelets, so it is possible to perform an infinite number of wavelet transforms depending on the kernel used. The kernel of the wavelet transform is usually denoted by the symbol Ψ, and it is simultaneously the time function t, the scale parameter a, and the translation parameter b. The parameter a shifts the wavelet spectrum in the frequency domain, while b parameter shifts it in time. Therefore, when the standard Fourier transform gives us ideal localization possibilities in the frequency domain but does not allow localization in the time domain, the use of the wavelet transform complements these deficiencies and enables us to perform full time-frequency analysis. Additionally, the time resolution of the wavelet transform is variable and depends on the wavelet frequency.

$$\widetilde{S}_{\Psi}(a,b) = \frac{1}{\sqrt{a}} \int_{-\infty}^{\infty} S(t)\, \Psi\!\left(\frac{t-b}{a}\right) dt \tag{1}$$

where a—scale parameter, b—time translation parameter, $S(t)$—analyzed signal, $\Psi((t-b)/a)$—the kernel of the wavelet transform, $S_\Psi(a, b)$—wavelet transform.

The scale parameter decides what pseudo-frequency will be represented by the wavelet. It always has values greater than zero, and they are inversely proportional to the pseudo-frequency of the wavelet (Figure 2). The $\frac{1}{\sqrt{a}}$ factor is responsible for wavelet normalization. This process is designed to keep the energy of the wavelet function constant in case of using different scale factors. The translation parameter (Figure 3) is responsible for the shift of the function, along with the tested signal. Changing the translation parameter moves function along the time axis [6].

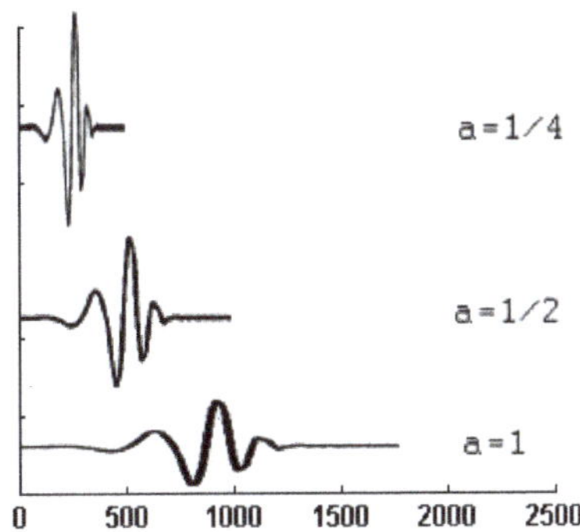

Figure 2. Example of the impact of changes in the scale factor on the wavelet [4].

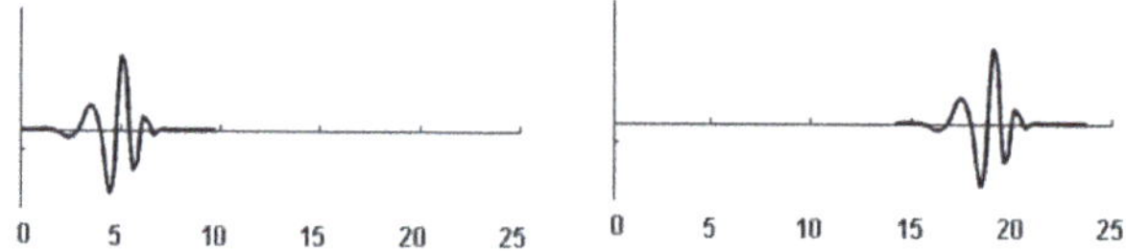

Figure 3. Effect of changing the translation parameter on wavelet [4].

Similarly to the method with Gaussian filtration, here also two complementary filters are used: high-pass (so called wavelet) and low-pass (scaling function). The wavelet analysis should be initiated by selecting a proper type of wavelet, and its shape must be as similar as possible to the analyzed signal to obtain the best results possible.

A number of studies have shown the use of different types of wavelets for signal analysis, of which the oldest and the simplest one is the Haar wavelet (HW)—originally called the Haar expansion. A significant disadvantage of the Haar wavelet is its non–differentiability. It is discontinuous, and therefore it is impossible to approximate continuous, smoothed functions. Additionally, the possibility of frequency localization is low in its case. Despite the fact that over the years, many other and more advanced wavelets have been invented, the HW is still willingly used due to its simplicity. In the papers [7,8], the analysis of signals with these functions was proposed in case of monitoring of the tool failure in machines and for analysis and optimization of drive systems. The best known, however, is a tool for actions on images, such as decomposition or compression, which is a typical use of discrete wavelets [9].

An extension of the Haar wavelet family is the Daubechies wavelet, whose characteristics are the compactness of the carrier with length $2N - 1$ (N is a wavelet row), a relatively simple form and an exact approximation of the function. What is more, the Daubechies Db1 wavelet corresponds to the Haar wavelet. Besides, as the row of the wavelet increases, the number of coefficients describing it increases too, and the waveform becomes smoother; as a result, it is often used as a tool for signal analysis. In [10], it was used to evaluate the chip formed during turning of S45C carbon steel, while in [11], to evaluate the vibrations generated during turning.

Wavelet analysis has been used in many studies of very diverse physical phenomena and in technical applications. The wavelet that appears most frequently in research related to signal analysis is the Morlet wavelet. This wavelet is a continuous wavelet characterized by good frequency resolution. It cannot be used for multiresolution analysis as it does not have a scaling function. However, it is an overt wavelet, so it can be represented by means of mathematical dependencies in the time domain. An example of the use of this wavelet can be found in the research [12] for the evaluation of the surface on the contact area between reinforcement and concrete. In [13], for the assessment of the phenomena occurring in the construction of motor vehicles during a collision.

Another wavelet, well described in research papers about signal analysis, used to assess the damage of the bearings operating in an electric motor is the Mexican hat wavelet [14]. Like the Morlet wavelet, it has no scaling function but has good frequency resolution. It comes from the Gaussian wavelet family—it is proportional to the function, which is the second derivative of the Gaussian probability density. Its name derives from a characteristic shape resembling a Mexican sombrero. Another example of the use of wavelet analysis as a tool to support the diagnosis of bearing damage and the assessment of its possibility of further use is the paper [15].

Analyzing, evaluating, and forecasting surface roughness is a very difficult issue. It is described by a large number of parameters, and their values are influenced by a number of different factors and input sizes to the machining process. Mathematical modeling can be used to predict surface roughness. This method is particularly important in case of difficulty in obtaining the actual surface. An example of modeling, where the predominating phenomena, leading to the obtaining of the machined surface, are physical and mechanical phenomena, is publication [16].

With a large number of factors affecting surface roughness, the optimal selection of machining parameters is very important. Such selection results in the appropriate, desired, and expected values of the roughness parameters. This is mainly the case in finishing machining, in particular, abrasive.

In order to achieve the assumed surface roughness in the manufacturing process, optimization can be carried out using a number of different methods and models. An example may be the ANOVA analysis [17] or the Monte Carlo method [18]. The ANOVA method is also used for multi-criteria optimization of the manufacturing process, in which the machinability indicator is not only surface roughness but also, for example, the vibration of the machining system [19]. Another example of optimization is the use of a hybrid WOA (Whale Optimization Algorithm) algorithm. Using this algorithm, a combination of machining parameters is obtained, which ensures maximum material removal speed and minimum surface roughness [20]. Neural networks and genetic algorithms are also used to predict machining effects and optimize the input parameters of the process [21].

Optimization methods give you the opportunity to select the input parameters for the process, which guarantees the surface roughness at a certain level. In many situations, especially in the case of surfaces with a special geometric structure, special application, or the surface of elements responsible, e.g., in aviation, a detailed analysis of these surfaces should be carried out. It can be assumed that a standard assessment using the roughness parameters obtained by Gaussian filtration is sufficient. On the other hand, since wavelength analysis is used to diagnose bearing damage, it can also be used to detect any deviations in the geometric structure of the machined surface.

That is why wavelet analysis, as an auxiliary tool for assessing surface roughness, has been studied by many scientists around the world. Such results can be found in, e.g., [22–24]. In [25], the Morlet wavelets and the Mexican hat were used to assess the condition of the turned surface of the C45 steel with a hardness of 55HRC. It was stated in this work that a Mexican hat wavelet gives information about the distribution of the roughness profile extremes and their values but does not allow for a precise evaluation of the wavelength. On the other hand, the Morlet wavelet allows the evaluation of the wavelengths of the profile components, but the information about their amplitudes is not accurate. In addition, it allows finding and assessing the intensity of disturbances occurring during machining, in particular disturbances in machining traces, and also to isolate major changes in line spacing along the entire length of the profile. In [11], 40 wavelets were tested in order to select the appropriate roughness for monitoring roughness during CNC (Computer numerical control) turning. The signals of the sensor measuring the vibrations were correlated with the measured surface roughness. Only a small number of mother wavelets showed good and relevant results. This shows that the mother wavelets should be selected according to the typology of the signal and the monitored factor. Moreover, it was found that the level of decomposition is very important. Too little or too much decomposition makes the results incorrect.

Attempts are made to use wavelet analysis to describe surface roughness also after erosion treatments. An example can be the work [26]. In the case of honed surfaces, the use of wavelet analysis to describe the surface condition was proposed in [27].

Attempts are being made to combine wavelet and fractal analysis as tools to describe the condition of machined surfaces [28]. In addition, advanced surface roughness analyses and geometric surface structures can be applied not only to the surfaces of machined components but also to cutting tools, more specifically, as analysis to support the assessment of tool wear. These can be very helpful tools, e.g., to assess the wear traces of tool coatings. The topic of a selection of suitable coatings for the machining of hard-to-machine materials and the assessment of their degree of wear is very important, and many scientific studies are raising this subject, e.g., in [29,30].

Despite a number of literature sources on the use of the wavelet transform to assess the surface condition of machined elements, this topic is still not fully understood and requires continuous research. Hence, the authors' interest in this subject.

3. Measurement Methodology

The aim of the work was to determine the possibility of using wavelet analysis to describe surface features obtained by abrasive treatment on the example of honing. The traces on such surfaces are random but also directional. Hence, their analysis may cause a number of difficulties. In order to achieve the goal of the work, it was necessary to perform tests according to a strictly defined plan:

- perform samples in a wide range of processing parameters so that it is possible to obtain characteristic traces on the surface but with different features,
- measure surfaces using the contact or non-contact method,
- perform initial filtration in the surface analysis program by leveling the measurement and removing shape errors,
- perform an analysis aimed at selecting a wavelet or a wavelet family,
- filter the surface with a Gaussian filter and selected wavelets, the effect of which will be the separation of surface roughness from waviness,
- compare the obtained 3D roughness and the calculated roughness parameters by Gaussian filtration and wavelets.

It was decided that the analyses would be performed on surfaces and not on roughness profiles. On the one hand, the easiest and most reliable way to select a wavelet for the roughness profile is when analyzing the surface roughness. On the other hand, one or even several profiles may not give enough information about the geometric structure of the surfaces. Hence, the authors decided to select the surfaces that were assembled by the software from 256 profiles.

In this research, four samples were tested, and their machining parameters are listed in Table 1. Each of the tests was performed with the same feeds and speeds but with a different granularity of the grinding stones. The abrasive grit size impacts the differences in the character and functional characteristics of the obtained surfaces.

Table 1. Machining parameters of the tested liners.

Liner Number	Granularity of Grinding Stone	Diameter and Length of the Liner (mm)	Spindle Speed (rpm)	Number of Hone Strokes Per Minute	Sparking Passes
1	55	Ø77,5; L-150	100	80	2
2	75 m	Ø77,5; L-150	100	80	2
3	85	Ø94,2; L-150	100	80	2
4	2 × 55; 4 × 75t	Ø94,2; L-150	100	80	2

Then, the state of the geometrical structure of the surface layer for each sample was measured using a Mitutoyo profilografometer, model SURFTEST SV-3200 (Figure 4), equipped with the MCubeMap Ultimate software. It is a contact measurement device, the performed measurements are fully automated, and it creates both surface roughness profiles and so-called contour maps (3D), allowing detailed analysis and processing of the obtained results.

Figure 4. Mitutoyo profilografometer SURFTEST SV-3200.

4. The Proposition of Selection of the Wavelet Type for the Assessment of the Flat-Top Structures with a Cross-Hatch Pattern

The compliance of the wavelet shape with the shape of the analyzed signal is an important factor of the wavelet analysis—the higher it is, the better results will be obtained as a result of the wavelet transform. In order to determine which wavelets will meet the above standards during the first phase of the analysis, a series of conversions of the measurements results of the liner number 1 (Table 1) was performed with the Daubechies wavelet using three different scaling sequences, and, for each applied row, at different filtering levels, respectively. The obtained results were compared with the result of the roughness profile filtrated using a standardized Gaussian filter. Therefore, the correctness of the results can be assumed. The Gaussian filtration with $\lambda c = 0.25$ mm was used, and the results for filtration with Gauss filter for sample 1 are presented in Figure 5.

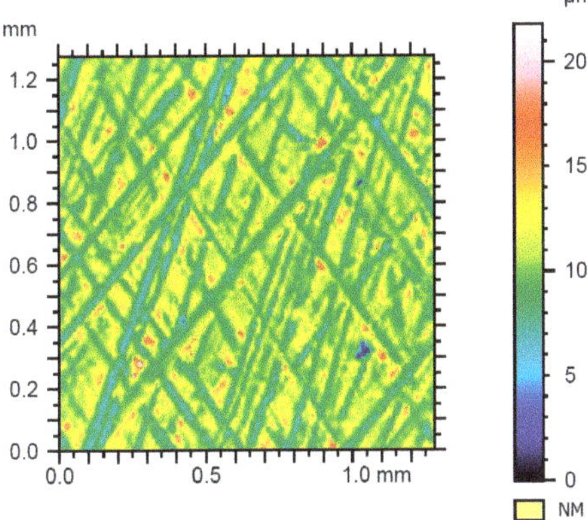

Figure 5. Measured roughness of sample no. 1 after filtration with Gaussian filter.

First, the Daubechies wavelet Db1 with the filtration levels of 1 and 6 was used to analyze the surface with a flat-top character. However, the nature of this wavelet (largely resembling the Haar wavelet) and analysis of the surface images obtained after its application led to the conclusion that the Db1 wavelet was not suitable for the analysis of this type of signal. This was confirmed by the results presented in Figures 6 and 7.

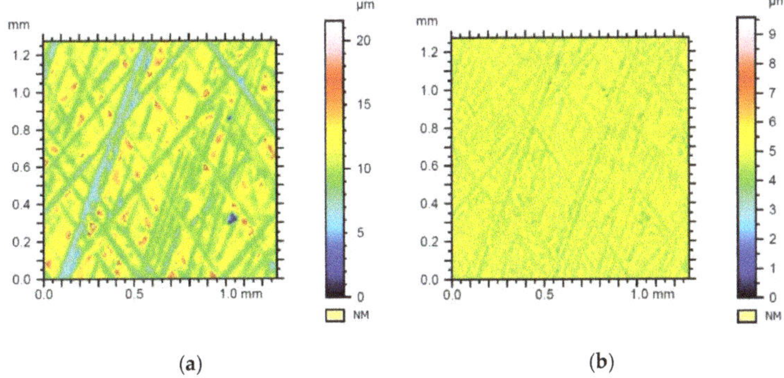

(a)   (b)

Figure 6. Results of filtration of the surface geometrical structure of sample 1 with Daubechies wavelet on the 1st row with the 1st level of filtration—(**a**) waviness, (**b**) roughness.

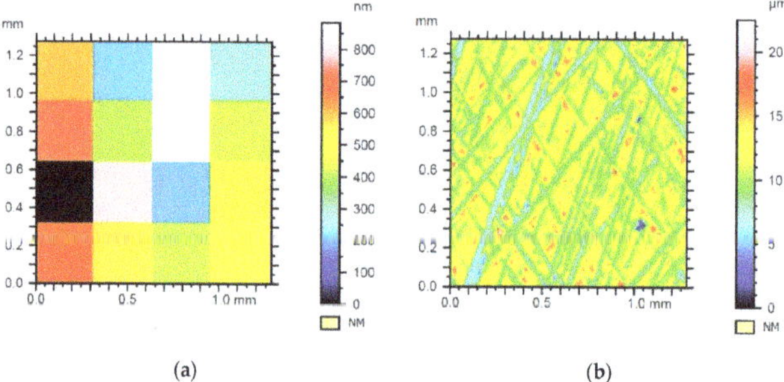

(a) (b)

Figure 7. Results of filtration of the surface geometrical structure of sample 1 with Daubechies wavelet on the 1st row with the 6th level of filtration—(**a**) waviness, (**b**) roughness.

In the next step of wavelet selection, the Daubechies wavelets on the 3rd and 6th row were used to describe the surface geometrical feature after honing with filtration level, respectively, from 1 to 6 and from 1 to 5. The results obtained for both Db3 and Db6 at low levels of filtration were rejected due to the large discrepancy with the result obtained with Gaussian filtration. For both tested wavelets, the obtained results were compared with the results obtained for Db3 and Db6, with the filtration level 5th being identical. Therefore, for further analyses, it was decided to use only Daubechies 6th row wavelet on at least the 5th level of the filtration. According to the authors of the study, only this level of filtration ensured complete filtering of the characteristic cross-intersecting traces from the waviness structure and their location on the roughness side. Figures 8–13 show the exemplary results of wavelet transformations with Db3 and Db6 wavelets on different levels of the filtration.

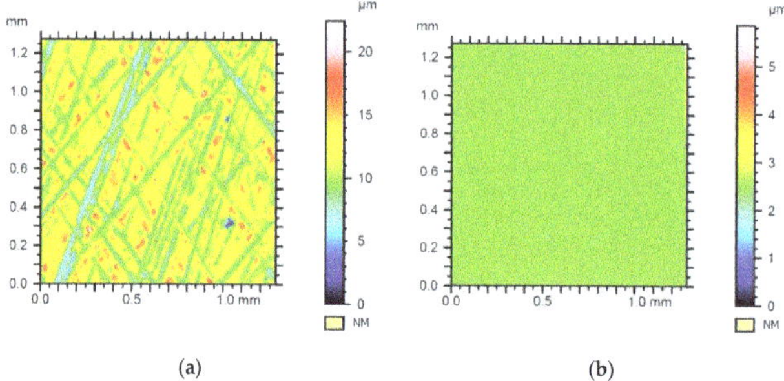

(a) (b)

Figure 8. Results of filtration of the surface geometrical structure of sample 1 with Daubechies wavelet of the 3rd row with the 1st level of filtration—(**a**) waviness, (**b**) roughness.

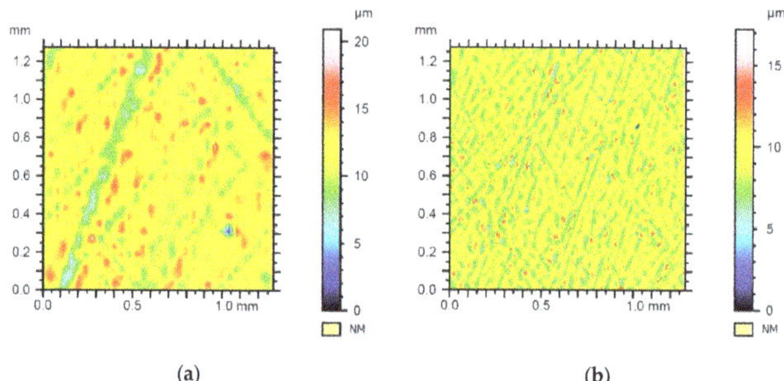

(a) (b)

Figure 9. Results of filtration of the surface geometrical structure of sample no. 1 with Daubechies wavelet of the 3rd row with the 3rd level of filtration—(**a**) waviness, (**b**) roughness.

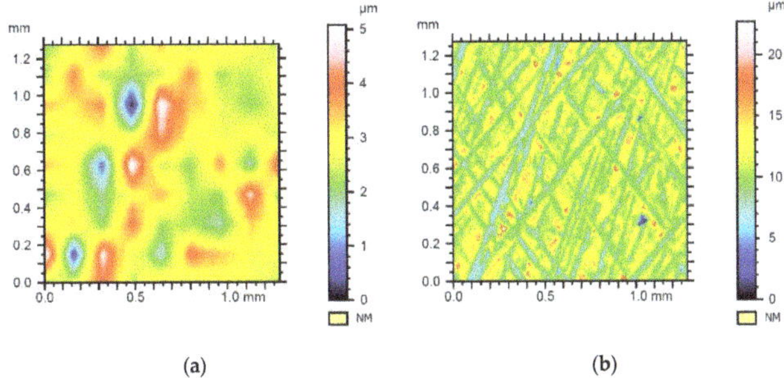

Figure 10. Results of filtration of the surface geometrical structure of sample no. 1 with Daubechies wavelet of the 3rd row on the 5th level of filtration—(**a**) waviness, (**b**) roughness.

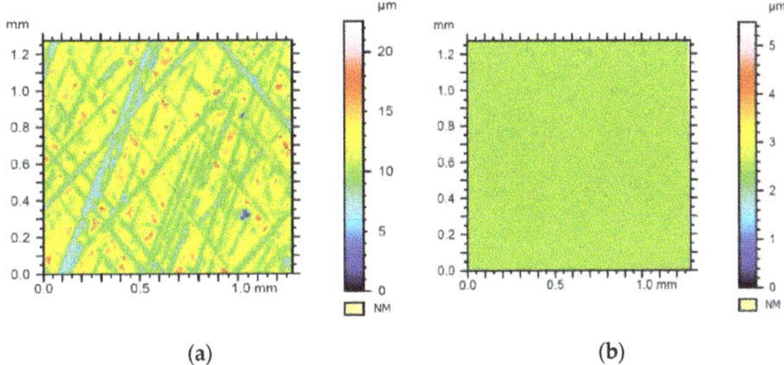

Figure 11. Results of filtration of the surface geometrical structure of sample no. 1 with Daubechies wavelet of the 6th row with the 1st level of filtration—(**a**) waviness, (**b**) roughness.

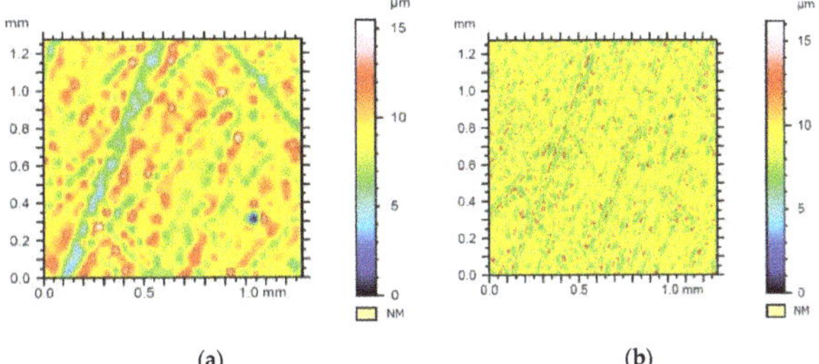

Figure 12. Results of filtration of the surface geometrical structure of sample no. 1 with Daubechies wavelet of the 6th row with the 3rd level of filtration—(**a**) waviness, (**b**) roughness.

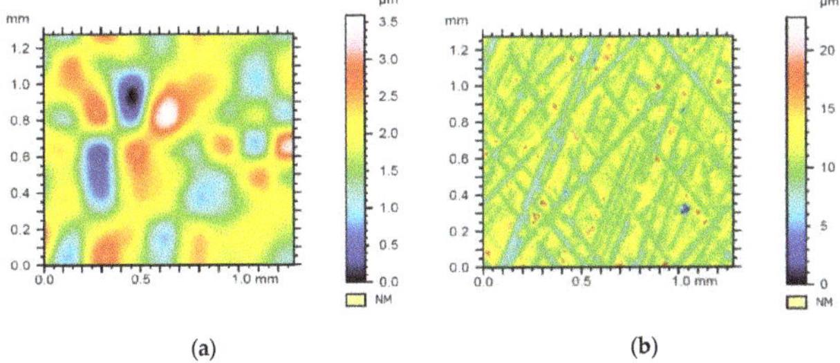

Figure 13. Results of filtration of the surface geometrical structure of sample no. 1 with Daubechies wavelet of the 6th row with the 5th level of filtration—(**a**) waviness, (**b**) roughness.

Based on subsequent analyses, two more wavelets were proposed for the assessment of the flat-top surfaces—the Morlet wavelet and the "Mexican hat" wavelet. Those wavelets had a similar shape, as previously selected Db6 wavelet. For these two wavelets, using at least the 5th level of the filtering was also proposed. However, during further research, it was found that for the Morlet wavelet, the highest possible setting of filtering was level 4th. Therefore, in this case, it was impossible to use the 5th level of filtration, which would correspond to the level of filtration used for the other wavelets.

5. Tests Results and Their Analysis

The measurements for all four samples, described in Table 1, were made using a square-shaped surface with a side length of 1.28 mm, with a resolution of 5 μm. The measurements were presented as a contour map of 256 linear measurements and 256 × 256 measurement points. Subsequently, they were presented as a contour map. The results for raw measurements for each sample were subjected to the following processes: leveling, shape removal, and filtration with Gauss filter with $\lambda c = 0.25$ mm or with wavelets selected in Section 4. Wavelets selected for research are shown in Figure 14. The initial surfaces for testing (samples 1–4) after their leveling and shape removing are shown in Figure 15.

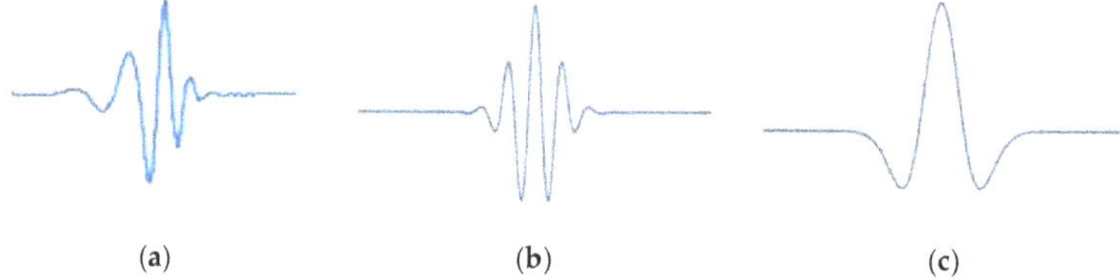

(a)   (b)   (c)

Figure 14. Wavelets used in the research: (**a**) Daubechies of 6th row, (**b**) Morlet, (**c**) Mexican hat [5].

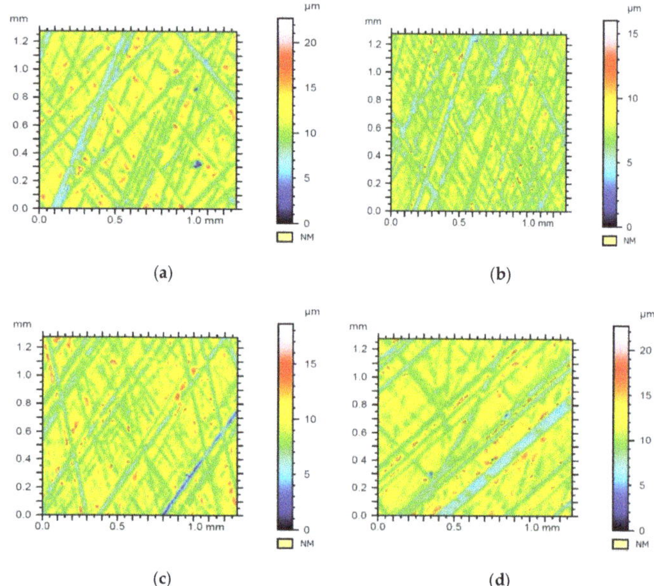

(a)   (b)

(c)   (d)

Figure 15. Surfaces of the samples tested with flat-top structures after their leveling and shape removal: (**a**) sample 1, (**b**) sample 2, (**c**) sample 3, (**d**) sample 4.

To assess the suitability of the individual wavelets used to describe geometrical features of the tested surfaces, the obtained results were compared with those achieved if the standardized Gaussian filter was used.

Additionally, the difference between contour maps of the surfaces obtained by Gaussian filtration and wavelet transformation would be calculated. This process would generate the root mean square error necessary to investigate the similarities between both surfaces and differences between the filtration tools; the lower the coefficient value, the higher the surface similarity. For each of the map differences, spatial parameters would be calculated and presented.

5.1. Sample 1

Sample 1 was honed with the 55 grit whetstone. The results for the Gaussian filtration compared to other preselected types of wavelets are shown in Figure 16. Table 2 shows the set of values for the selected 3D spatial parameters for surfaces shown in Figure 17.

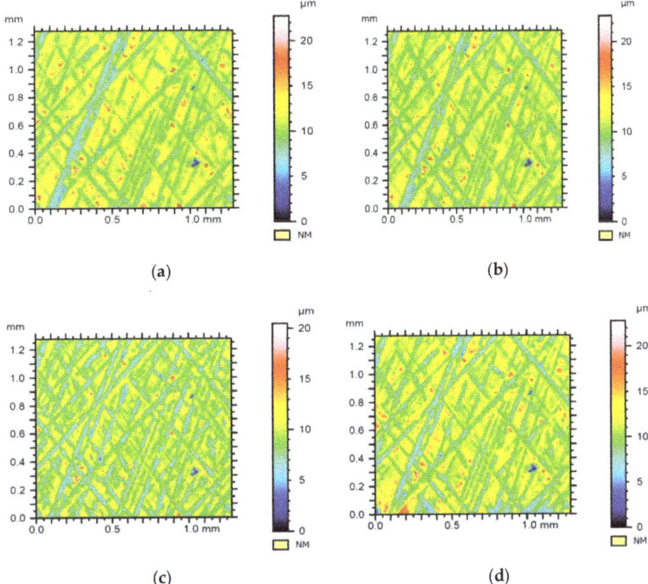

Figure 16. The surface of sample 1 with a flat-top structure after leveling and shape removal and application of (**a**) Gaussian filtration, (**b**) Daubechies wavelet, (**c**) Morlet wavelet, (**d**) Mexican hat wavelet.

Table 2. Selected special parameters calculated for surfaces shown in Figure 17.

Parameter	Gauss-Db6	Gauss-Morlet	Gauss-Mexican Hat
Sq [µm]	0.34	0.58	0.678
Ssk	−0.27	−0.09	0.33
Sku	3.3	3.47	12.5
Sp [µm]	1.09	2.27	7.23
Sv [µm]	1.31	2.79	4.31
Sz [µm]	2.35	5.21	10.9
Sa [µm]	0.26	0.49	0.39
Vm [mm^3/mm^2]	1.5×10^{-5}	3.11×10^{-5}	6.3×10^{-5}
Vv [mm^3/mm^2]	0.000475	0.00082	0.000761
Vmp [mm^3/mm^2]	1.5×10^{-5}	3.11×10^{-5}	6.3×10^{-5}
Vmc [mm^3/mm^2]	0.000298	0.00057	0.000398
Vvc [mm^3/mm^2]	0.00038	0.000726	0.00062
Vvv [mm^3/mm^2]	4.57×10^{-5}	7.4×10^{-5}	8.6×10^{-5}
Spd [1/mm^2]	13	27.8	2.38
Spc [1/mm]	0.314	1.06	0.338

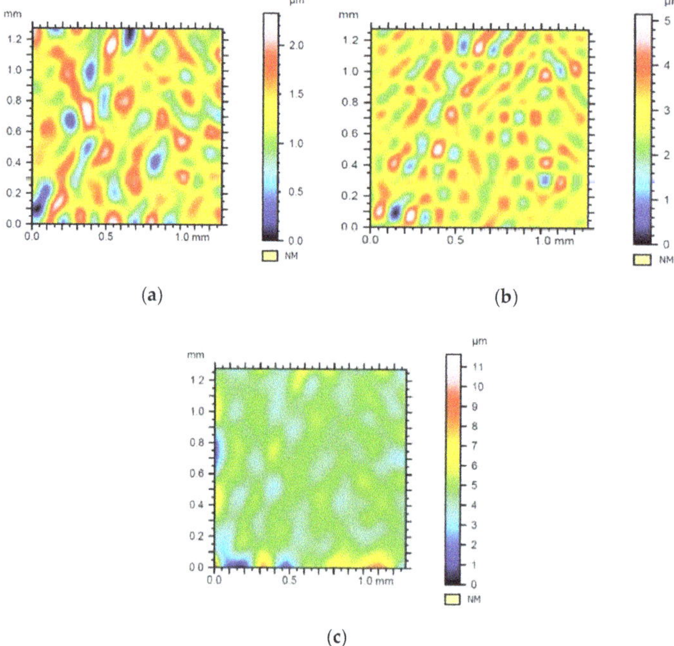

Figure 17. Surfaces created as a result of subtracting structure obtained with a wavelet from the structure after applying Gauss filtering: (**a**) Daubechies wavelet, (**b**) Morlet wavelet, (**c**) Mexican hat wavelet. The root mean square error: (**a**) 0.36 μm, (**b**) 0.62 μm, (**c**) 0.68 μm.

Figure 17 shows the differences in contour maps between the surfaces obtained after the application of the Gaussian filter and surfaces generated after other wavelets were used.

As can be seen in Figure 16a, based on low values of *Sp* and *Sv* parameters in comparison to the *Sa* parameter value (Table 2), it could be determined that the surface topography was quite regular. The *Sku* parameter value also confirmed this regularity. Therefore, in this case, the analysis of the parameters did not provide a clear difference between the wavelet transform with Db6 wavelet and Gauss filtering. The root mean square error value of 0.36 μm determined a fairly high degree of similarity for both surfaces. However, when analyzing Figure 17a, it can be seen that the area where the difference between Gaussian filtration and Db6 wavelet appeared (marked in red) overlapped with the edge of the furrows on the surface of the tested sample. Since the furrows were places where the profile height changed rapidly with respect to the material core, it could be concluded that wavelet transform with Db6 wavelet enhanced the zones of the greatest signal gradient.

In the case of Figure 16b, the root mean square error value was twice as high, which means the similarity between reference surfaces was significantly smaller. The *Sku* value was similar to the one in the first case, but the value of the Sa parameter had almost doubled (Table 2). Therefore, the examined surface was regular. However, the average height of the material core was higher, and the surface topography had greater peaks and valleys, as evidenced by *Sp* and *Sv* parameters. In this case, high values of *Spd* and *Spc* parameters deserved special attention. They, respectively, informed us about a large number of peaks (and their density) and about their average curvative. When looking at the contour map in Figure 17b, it can be seen that red zones were showing large differences between reference surfaces located on the borders of furrows, in places where the height change occurred. This led to the conclusion that the Morlet wavelet transformation allowed finding zones where local changes of signal occurred.

The 3rd contour map (Figure 17c) was the result of the surface difference obtained using Gaussian filtration and the Mexican hat wavelet transform and had a very similar value of the root mean square and *Sa* parameter to Figure 17b. However, in contrast to the previous contour maps, the values of *Sku, Sp, Sv* parameters were much higher. This means that with the general regularity of the core topography on the contour map, there were single large valleys and peaks, rarely scattered over the entire surface of the sample, indicated by the low value of the *Spd* parameter. This means that the "Mexican hat" wavelet primarily filtered zones in which the signal extremes were located.

5.2. Sample 2

Sample 2 was honed with 75 m grit whetstone. The use of a whetstone, with a smaller granularity and abrasive grains, stacked closer together, gave more furrows less roughness and a surface with more regular topography. As a result, this affected the outcome of the wavelet transform. The low root mean square error values obtained during the surface subtraction operation of 0.18 μm, 0.27 μm, and 0.31 μm indicated a high degree of similarity of the surfaces obtained when using selected wavelets to the surfaces filtered with standardized Gaussian filter which are shown in Figure 18. Therefore, in the case of surfaces with more regular topography and lower roughness values, the effect of selected wavelet transform was similar to the use of Gaussian filtering, and the importance of the method itself decreased.

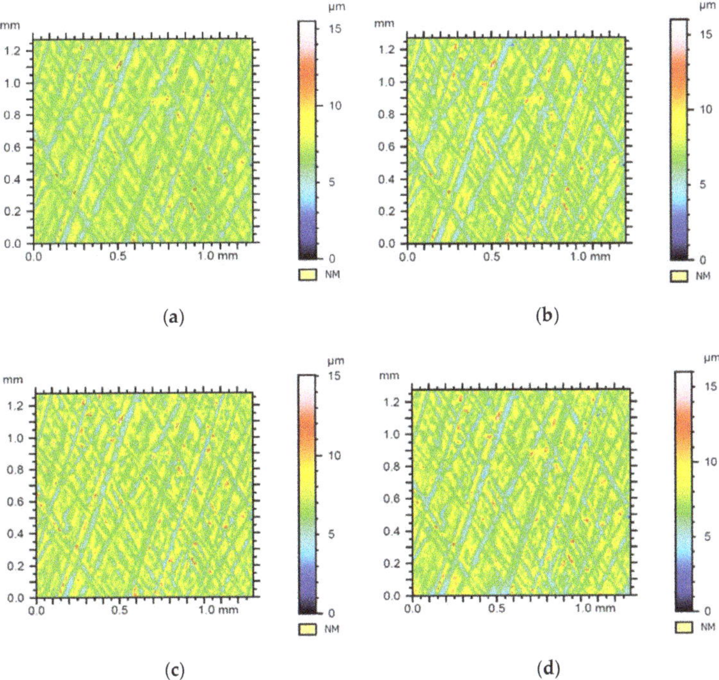

Figure 18. The surfaces of sample no. 2 with a flat-top structure after its leveling, shape removing, and applying: (**a**) Gauss filtering, (**b**) Daubechies wavelet, (**c**) Morlet wavelet, (**d**) Mexican hat wavelet.

Despite the greater similarity of contour maps of surface differences, for sample no. 2, the same trends were maintained, as in the case of sample 1. Similarly to sample 1, in sample 2, the Db6 wavelet generated the lowest values of roughness height (amplitude) parameters *Sp, Sv,* and *Sa* (Table 3). However, for sample 2, it was more difficult to see the visual relationships described for sample 1. This further confirmed that the use of selected wavelets was less important for the detection of features in case of similar surfaces. The high value of *Spd* parameter (Figure 19b) and high values of parameters *Sp, Sv,* and *Sa* with a low value of *Spd* parameter (Figure 19c) confirmed the hypothesis for using Morlet wavelet to find individual peaks and valleys and Mexican hat wavelet to filter entire zones, where extreme signals were present.

Table 3. Selected spatial parameters, calculated for surfaces shown in Figure 19.

Parameter	Gauss-Db6	Gauss-Morlet	Gauss-Mexican Hat
Sq [µm]	0.18	0.27	0.328
Ssk	−0.16	−0.02	−0.533
Sku	4.4	3.07	20.5
Sp [µm]	0.88	0.95	2.33
Sv [µm]	0.69	1.03	4.71
Sz [µm]	1.55	2.11	7.29
Sa [µm]	0.131	0.21	0.169
Vm [mm^3/mm^2]	9×10^{-6}	1.21×10^{-5}	2.3×10^{-5}
Vv [mm^3/mm^2]	0.000475	0.000377	0.000261
Vmp [mm^3/mm^2]	9×10^{-6}	1.21×10^{-5}	2.3×10^{-5}
Vmc [mm^3/mm^2]	0.000138	0.000257	0.000188
Vvc [mm^3/mm^2]	0.00018	0.000326	0.000252
Vvv [mm^3/mm^2]	2.5×10^{-5}	3.09×10^{-5}	4.6×10^{-5}
Spd [1/mm^2]	11	26.8	1.31
Spc [1/mm]	0.14	0.46	0.233

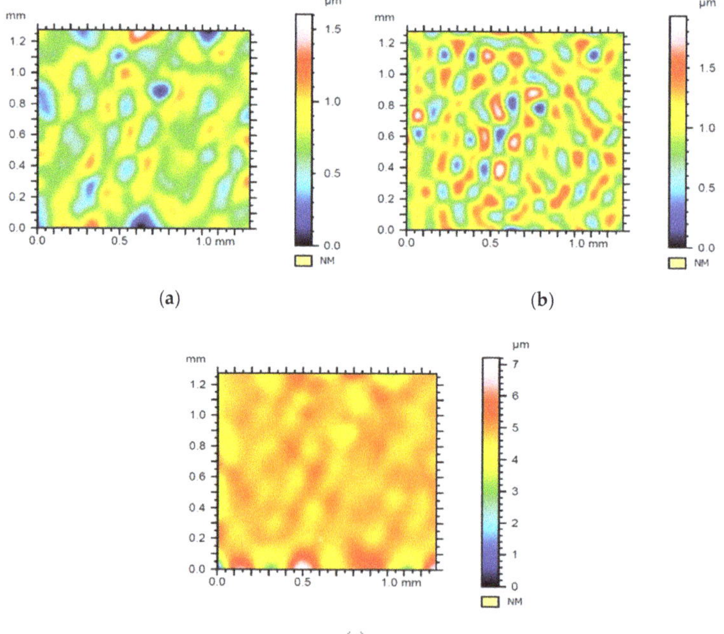

(a)　　　　(b)

(c)

Figure 19. Surfaces created as a result of subtracting the structure obtained with wavelet from the structure after using Gaussian filtering: (**a**) Daubechies wavelet, (**b**) Morlet wavelet, (**c**) Mexican hat wavelet. The root mean square error: (**a**) 0.18 µm, (**b**) 0.27 µm, (**c**) 0.31 µm.

5.3. Sample 3

Sample 3 was machined with the 85 grit whetstone. The surface topography for sample 3 was less regular than the one for sample 2, despite using smaller abrasive grains of the whetstone. A significant number of single peaks scattered over the entire sample surface with one furrow with a significant depth located in the right, and the lower corner is visible in Figure 20. This might indicate the additional point flashes of material formed along with deeper scratches.

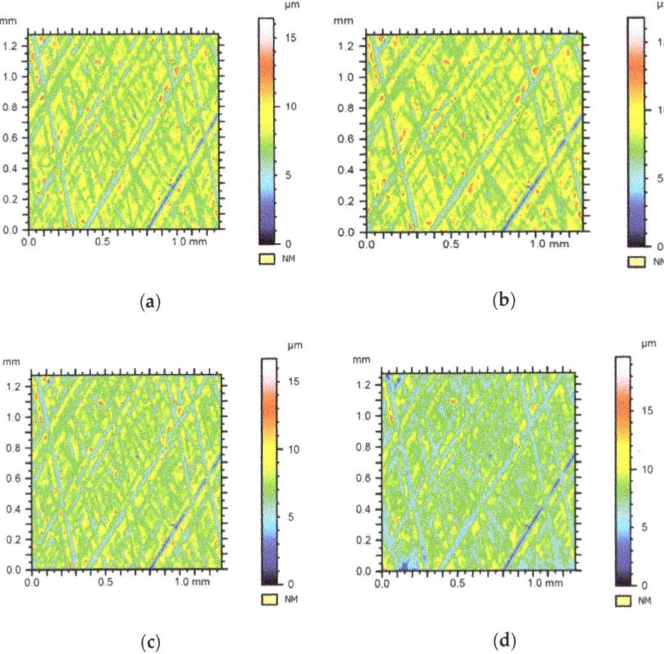

Figure 20. The surface of sample no. 3 with a flat-top structure after leveling, shape removing, and applying: (**a**) Gaussian filter, (**b**) Daubechies wavelet, (**c**) Morlet wavelet, (**d**) Mexican hat wavelet.

The results for the root mean square errors for sample 3 were higher than for sample 2. Hence, it could be concluded that the selected types of wavelets might be well used for assessing the uniformity of scratches depth made during honing with 85 grit whetstone and uniformity of the material machined or when the whetstones generating such errors were used.

Further analysis of the results for sample 3 revealed that only transformation with a Mexican hat wavelet showed differences in surfaces when compared to samples 1 and 2 (Figure 21). A similar tendency of the *Sku*, *Sp*, and *Sv* parameters was observed when compared to previous samples. However, the value of the *Spd* parameter was 0; therefore, the value of the Spc parameter could not be calculated (Table 4). This was due to the location of extremes (both single point peaks and several deeper scratches visible on all measured surface). Mexican hat wavelet emphasized the extreme points (especially towards the valleys) and made it possible to detect the additional point zone of the material valley, which is visible in the left low corner of Figure 20d.

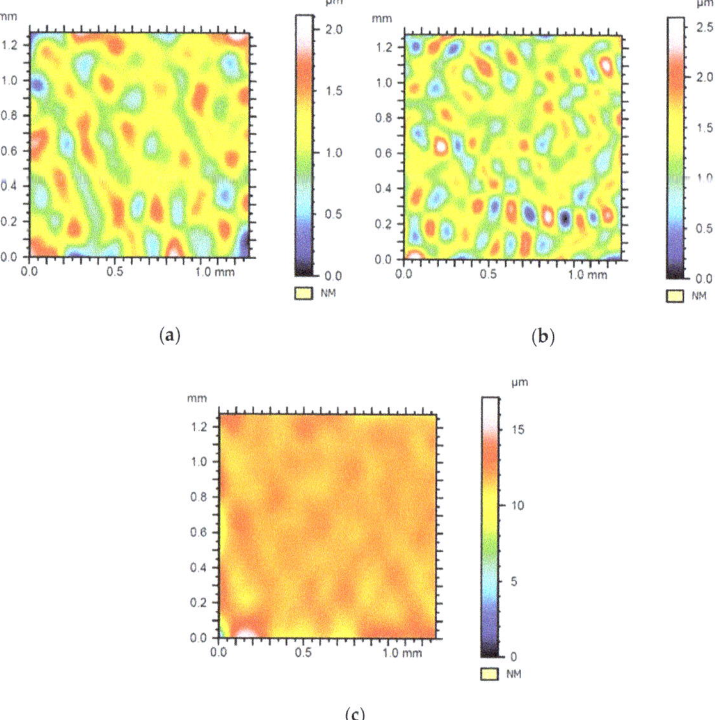

Figure 21. The surface was created as a result of subtracting the structure obtained with the wavelet from the structure after applying Gaussian filtration: (**a**) Daubechies wavelet 6, (**b**) Morlet wavelet, (**c**) Mexican hat wavelet. The root mean square error: (**a**) 0.28 μm, (**b**) 0.32 μm, (**c**) 0.53 μm.

Table 4. Selected spatial parameters, calculated for surfaces shown in Figure 21.

Parameter	Gauss-Db6	Gauss-Morlet	Gauss-Mexican Hat
Sq [μm]	0.28	0.327	0.528
Ssk	−0.1	0.052	−0.95
Sku	3.4	3.37	48.5
Sp [μm]	1.08	1.35	5.33
Sv [μm]	1.09	1.23	11.91
Sz [μm]	2.15	2.51	16.3
Sa [μm]	0.231	0.261	0.316
Vm [mm³/mm²]	1.23×10^{-5}	1.9×10^{-5}	5.5×10^{-5}
Vv [mm³/mm²]	0.000375	0.000437	0.000526
Vmp [mm³/mm²]	1.23×10^{-5}	1.9×10^{-5}	5.5×10^{-5}
Vmc [mm³/mm²]	0.000238	0.000277	0.00026
Vvc [mm³/mm²]	0.000318	0.00036	0.000452
Vvv [mm³/mm²]	3.15×10^{-5}	4.09×10^{-5}	5.86×10^{-5}
Spd [1/mm²]	11.9	24.8	0
Spc [1/mm]	0.21	0.54	*****

***** means that it was impossible to calculate a given parameter.

5.4. Sample 4

This chapter reviews sample no. 4, which was machined with two 55 grit whetstones and four 75t grit whetstones. Figure 22a–d show results of filtration with the Gaussian method compared to the wavelet transform with three types of wavelets. Table 5 shows a set of values of selected 3D spatial parameters for surfaces presented in Figure 23a–c. This figure presents contour maps of differences between the surfaces obtained after applying Gaussian filtering and surfaces, for generating which, the wavelets of tested types were used.

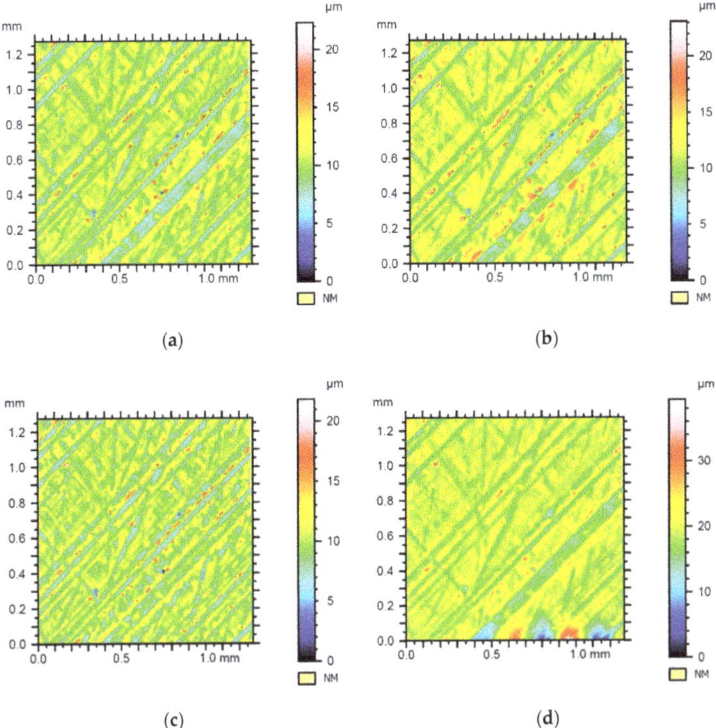

Figure 22. Surfaces of sample no. 4 with flat-top structure, after leveling, shape removing, and applying: (**a**) Gaussian filter, (**b**) Daubechies wavelet, (**c**) Morlet wavelet, (**d**) Mexican hat wavelet.

Table 5. Selected spatial parameters calculated for surfaces shown in Figure 23.

Parameter	Gauss-Db6	Gauss-Morlet	Gauss-Mexican Hat
Sq [µm]	0.48	0.74	1.63
Ssk	−0.091	−0.089	0.53
Sku	3.16	4.27	33.5
Sp [µm]	1.58	2.35	15.95
Sv [µm]	1.69	2.23	17.93
Sz [µm]	3.25	5.11	34.9
Sa [µm]	0.328	0.55	0.676
Vm [mm³/mm²]	2.03×10^{-5}	3.9×10^{-5}	0.000179
Vv [mm³/mm²]	0.000675	0.000877	0.000865
Vmp [mm³/mm²]	2.03×10^{-5}	3.9×10^{-5}	0.000179
Vmc [mm³/mm²]	0.00032	0.000588	0.000371
Vvc [mm³/mm²]	0.00048	0.00086	0.000583
Vvv [mm³/mm²]	5.85×10^{-5}	9.19×10^{-5}	0.000223
Spd [1/mm²]	15.2	33.4	0
Spc [1/mm]	0.39	1.13	*****

***** means that it was not possible to calculate the given parameters.

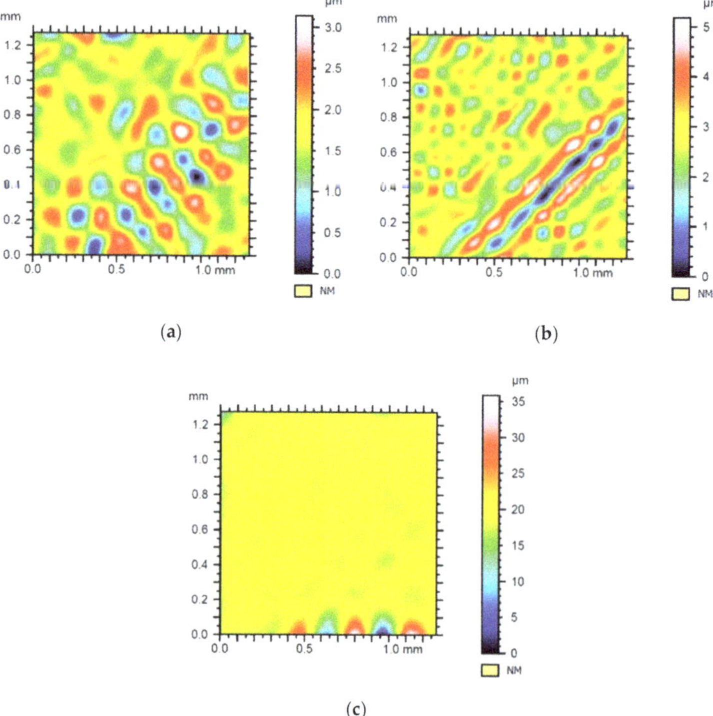

Figure 23. The surface was created as a result of subtracting the structure obtained with the wavelet with the wavelet from the structure after applying Gaussian filtration (**a**) Daubechies wavelet, (**b**) Morlet wavelet, (**c**) Mexican hat wavelet. The root mean square error: (**a**) 0.48 μm, (**b**) 0.72 μm, (**c**) 1.63 μm.

Figure 22 displays the differences in widths and the arrangement of furrows obtained when using different types of whetstones in one tool. The contour maps presented in Figure 23 show the same trends as previous samples and confirm initial conclusions. Both contour maps, shown in Figures 21c and 23c, share a number of key features. The use of the Mexican hat wavelet led to the detection of places where extreme points were placed (visible in the lower part of Figure 23c). Their detection made it impossible for the program to calculate the values of *Spd* and *Spc* parameters. The occurrence of extreme places near the border of the measurement field might indicate that the Mexican hat wavelet could be particularly useful for the detection of errors in these very places of the assessed images. The Gaussian filter (due to its characteristics) introduced, particularly, many inaccuracies for this kind of sample.

6. Summary and Conclusions

The paper presents the results of research and analyses of the application of the wavelet transform to assess the surface condition of honed cylindrical sleeves. The advantage of the presented method is the possibility of quick analysis, which can be performed in one software, which is often available with a measuring device. In the course of the research, it is found that the application of various types of wavelets for the profile analysis allows us to identify and observe different properties, such as profile roughness or the fragments of the surfaces measured. The appropriate wavelet will emphasize the feature we want to highlight. The Daubechies wavelet allows for the observation of places with a high signal gradient and highlights the areas where the signal value changes significantly. While the Morlet wavelet shows the zones of local changes of a signal, and the Mexican hat wavelet exposes points of local extremes.

As seen above, it has many advantages. However, the findings in this report also show a number of disadvantages, which are shown when samples with a high degree of regularity and low roughness parameters are used. Therefore, the above recommendations for the use of individual types of wavelets best works for sample 1, which is characterized by a high degree of profile irregularity. Subsequent samples, especially sample 2, which is the most regular of all and whose roughness profile does not have many defects, shows that the use of wavelet analysis (in the scope of evaluated wavelets) is not applicable, or its application is limited due to the significant similarity of the surfaces obtained after filtration with standardized Gaussian filter and analyzed wavelets.

In conclusion, a wavelet transform is a useful tool, supporting the analyses of the roughness profile in terms of searching for signal properties other than the classical methods of its filtering. The major limitation of this method is the selection of the appropriate wavelet for the assessment and observation of individual groups of features occurring on the real surfaces. The second limitation is the fact that wavelet transforms, using each of the wavelets applied in this paper, do not meet the expectations when profiles or surfaces with a high degree of regularity and low values of peaks or valleys to the average surface are analyzed. Therefore, there is a selection of another wavelet of which properties and shape will be a better fit for the nature of the analyzed signal.

The next stage of the research will be the analysis of the possibility of using the wavelet transform to evaluate the geometrical structures of surfaces with different properties. The result of extensive research should be a set of recommendations on what types of wavelets should be used to describe a specific type of surface geometric structure. The development of such a set of recommendations will significantly speed up the entire analysis.

Author Contributions: Conceptualization, M.K. and P.K.; methodology, M.K.; software, P.K.; validation, M.K., P.K. and M.W.; formal analysis, P.K.; investigation, M.K.; resources, M.K., P.K., M.W.; data curation, M.W.; writing—original draft preparation, M.K.; writing—review and editing, M.W.; visualization, M.W.; supervision, P.K.; All authors have read and agreed to the published version of the manuscript.

Funding: This research received no external funding.

Conflicts of Interest: The authors declare no conflict of interest.

References

1. Wieczorowski, M. *Theoretical Foundations of the Spatial Analysis of Surface Irregularities*; Publishing house of Wrocław Board FSNT NOT: Wrocław, Poland, 2013; pp. 7–34. (In Polish)
2. Chmielik, I.; Czarnecki, H. *Surface Topography Analysis in 3d System Using Filtration in Rectangle Format*; Mechanik: Warszawa, Poland, 2016; Volume 7, pp. 664–665.
3. Sundararajan, D. *Discrete Wavelet Transform: A Signal Processing Approach*; John Wiley & Sons: Singapore, 2015.
4. Białasiewicz, J. *Wavelets and Approximations*; Wydawnictwa Naukowo-Techniczne: Warszawa, Poland, 2004. (In Polish)
5. Daubechies, I. *Ten Lectures on Wavelets*; Society for Industrial and Applied Mathematics: Philadelphia, PA, USA, 1997.
6. Glabisz, W. *Package Wavelet Analysis in Problems of Mechanics*; Oficyna Wydawnicza Politechniki Wrocławskiej: Wroclaw, Poland, 2004. (In Polish)
7. Lee, B.Y.; Tarng, Y.S. Application of the Discrete Wavalet Transform to the Monitoring of Tool Failure in End Milling Using the Spindle Motor Current. *Int. J. Adv. Manuf. Technol.* **1999**, *15*, 238–243. [CrossRef]
8. Kosicka, M. *Possibilities of Using Haar Wavelets for Analysis and Optimization of Drive Systems (In Polish)*; Prace Instytutu Elektrotechniki: Warszawa, Poland, 2002; Volume 214, pp. 99–116.
9. Zawada-Tomkiewicz, A. Analysis of the Image of the Machined Surface for the Purpose of Estimating the Parameters of this Surface. *Acta Mech. Autom. Białystok* **2007**, *1*, 79–84. (In Polish)
10. Tangjitsitcharoen, S.; Haruetai, L. Hybrid monitoring of chip formation and straightness in CNC turning by utilizing Daubechies Wavelet Transform. *Procedia Manuf.* **2018**, *25*, 279–286. [CrossRef]
11. Garcia Plaza, E.; Nunez Lopez, P.J. Application of the wavelet packet transform to vibration signals for Surface roughness monitoring in CNC turning operations. *Mech. Syst. Signal Process.* **2018**, *98*, 902–919. [CrossRef]

12. Chang, C.-C.; Yu, C.-P.; Lin, Y. Distinction Between Crack Echoes and Rebar Echoes Based on Morlet Wavelet Transform of Impact Echo Signals. *NDT E Int.* **2019**, 108. [CrossRef]

13. Karimi, H.R.; Pawlus, W.; Robbersmyr, K.G. Signal reconstruction, modeling and simulation of vehicle full-scale crash test based on Morlet wavelets. *Neurocomputing* **2012**, *93*, 88–99. [CrossRef]

14. Liu, W.Y.; Han, J.G. The optimal Mexican hat wavelet filter denoisig method based on cross-validation method. *Neurocomputing* **2013**, *108*, 31–35. [CrossRef]

15. Nikravesh, S.Y. Intelligent Fault Diagnosis of Bearings Based on Energy Levels in Frequency Bands Using Wavelet and Support Vector Machines (SVM). *J. Manuf. Mater. Process.* **2019**, *3*, 11.

16. Brown, I.; Schoop, J. An Iterative Size Effect Model of Surface Generation in Finish Machining. *J. Manuf. Mater. Process.* **2020**, *4*, 0063.

17. Ezeddini, S.; Boujelbene, M.; Bayraktar, E.; Ben Salem, S. Optimization of the Surface Roughness Parameters of Ti–Al Intermetallic Based Composite Machined by Wire Electrical Discharge Machining. *Coatings* **2020**, *10*, 900. [CrossRef]

18. Kowalczyk, M.; Tomczyk, K. Procedure for Determining the Uncertainties in the Modeling of Surface Roughness in the Turning of NiTi Alloys Using the Monte Carlo Method. *Materials* **2020**, *13*, 4338. [CrossRef] [PubMed]

19. Kuntoglu, M. Modeling of Cutting Parameters and Tool Geometry for Multi-Criteria Optimization of Surface Roughness and Vibration via Response Surface Methodology in Turning of AISI 5140 Steel. *Materials* **2020**, *13*, 4242. [CrossRef]

20. Tanvir, M.H. Multi-Objective Optimization of Turning Operation of Stainless Steel Using a Hybrid Whale Optimization Algorithm. *J. Manuf. Mater. Process.* **2020**, *4*, 64. [CrossRef]

21. Lalwani, V. Response Surface Methodology and Artificial Neural Network-Based Models for Predicting Performance of Wire Electrical Discharge Machining of Inconel 718 Alloy. *J. Manuf. Mater. Process.* **2020**, *4*, 44.

22. Josso, B.; Burton, D.; Lalor, M.J. Frequency normalised wavelet transform for Surface roughness analysis and characterisation. *Wear* **2002**, *252*, 491–500. [CrossRef]

23. Gogolewski, D. Influence of the edge effect on the wavelet analysis process. *Measurement* **2020**, *152*, 107314. [CrossRef]

24. Rishi Pahuja, M. Ramulu Study of surface topography in Abrasive Water Jet machining of carbonfoam and morphological characterization using Discrete Wavelet Transform. *J. Mater. Process. Tech.* **2019**, *273*, 116249. [CrossRef]

25. Grzesik, W.; Brol, S. Wavelet and fractal approach to surface roughness characterization after finish turning of different workpiece materials. *J. Mater. Process. Technol.* **2009**, *209*, 2522–2531. [CrossRef]

26. Mahashar Ali, J.; Siddhi Jailani, H.; Murugan, M. Surface roughness evaluation of electrical discharge machined surfaces using wavelet transform of speckle line images. *Measurement* **2020**, *149*, 107029. [CrossRef]

27. Sabri, L.; Mezghani, S.; Mansori, M.; Zahouani, H. Multiscale study of finish-honing process in pass production of cylinder liner. *Wear* **2011**, *271*, 509–513. [CrossRef]

28. Li, G.; Zhang, K.; Gong, J.; Jin, X. Calculation method for fractal characteristics of machining topography surface based of wavelet transform. *Procedia CIRP* **2019**, *79*, 500–504. [CrossRef]

29. Silva, F.J. Machining GX2CrNiMoN26-7-4 DSS Alloy: Wear Analysis of TiAlN and TiCN/Al2O3/TiN Coated Carbide Tools Behavior in Rough End Milling Operations. *Coatings* **2019**, *9*, 392. [CrossRef]

30. Vitor, F.C.; Francisco, J.G. Recent Advances on Coated Milling Tool Technology—A Comprehensive Review. *Coatings* **2020**, *10*, 235.

Publisher's Note: MDPI stays neutral with regard to jurisdictional claims in published maps and institutional affiliations.

Article

The Possibilities of Improving the Fatigue Durability of the Ship Propeller Shaft by Burnishing Process

Stefan Dzionk *, Włodzimierz Przybylski and Bogdan Ścibiorski [1]

Department of Manufacturing and Production Engineering, Faculty of Mechanical Engineering, Gdansk University of Technology, 80-233 GDAŃSK, Poland; wprzybyl@pg.edu.pl (W.P.); bogdan.scibiorski@pg.edu.pl (B.Ś.)
* Correspondence: stefan.dzionk@pg.edu.pl; Tel.: +48-58-347-12-82

Received: 31 August 2020; Accepted: 15 October 2020; Published: 18 October 2020

Abstract: Heavily loaded structural elements operating in a corrosive environment are usually quickly destroyed. An example of such an element is a ship propeller operating in a seawater environment. This research presents a fatigue resistance test performed on elements operating in seawater. Different processing parameters applied on the samples in particular were compared with the specimens whose surface had been burnished differently and they were compared to specimens with a grinded surface. The research shows that the structural elements whose surface has been burnished can have up to 30% higher fatigue strength in a seawater environment than the elements whose surface has been grinded. During burnishing, an important feature of the process is the degree of cold rolling of the material. The resistance of the component to fatigue loads increases only to a certain level with increasing the degree of the cold rolling. Further increasing the degree of cold rolling reduces the fatigue strength. Introducing additional stresses in the components (e.g., assembly stresses) reduces the fatigue strength of this component in operation and these additional stresses should be accounted for while planning the degree of the cold rolling value. A device that allows for simultaneous turning and shaft burnishing with high slenderness is presented in the appendix of this article. This device can be connected to the computerized numerical control system and executed automatic process according to the machining program; this solution reduces the number of operations and cost in the process.

Keywords: burnishing process; fatigue strength; ship propeller; surface layer; surface processing

---

1. Introduction

The propeller drive is currently a basic way of propulsion for ships and other watercrafts (fish boats, yachts, et al.). The characteristic element of this drive is the propeller placed on the shaft extending beyond the hull outline. The propeller elements work in an environment of seawater and they transmit heavy loads; therefore, they must be carefully designed and produced. They are important parts used for the safety of the ship. Unfortunately, the detailed causes of ship accidents are limited in statistical databases due to the interests of ship-owners. On the other hand, Bureau Veritas [1] presents data for the number of fatigue damages in the ship propeller shaft and in recent years, more similar failures have been recorded (in this case, there is a lack of confirmed data, while it is estimated at about several dozen per year) [2,3]. This article presents proposals for a machining method for the ship propeller shaft where the surface layer is smoothed and strengthened by the burnishing process. This process increases fatigue strength and corrosion resistance, which affects the durability and life of the propeller shaft and applies to other structural elements operating in a harsh environment.

1.1. Loads of Propeller Shaft

The failures of ship screw engine shafting are the result of shaft work in particularly harsh environmental conditions. There are, among others, the dynamic load of the shaft as well as the seawater environment. The main shaft load forces are:

- The torsional torque transmitted by the ship's propulsion system;
- Compressive stress induced by the driving force of the propeller;
- Bending torque of the shaft from screw gravity force;
- The loads as a result of vibration of the powertrain system.

These loads are, among others, a reason to create phenomena to expedite the destruction processes of operating elements of the propeller system. There are, among others, pitting corrosion, fretting corrosion, and fatigue cracks on the shaft surface. The scheme of occurrence of these phenomena at the screw shaft joints is shown in Figure 1.

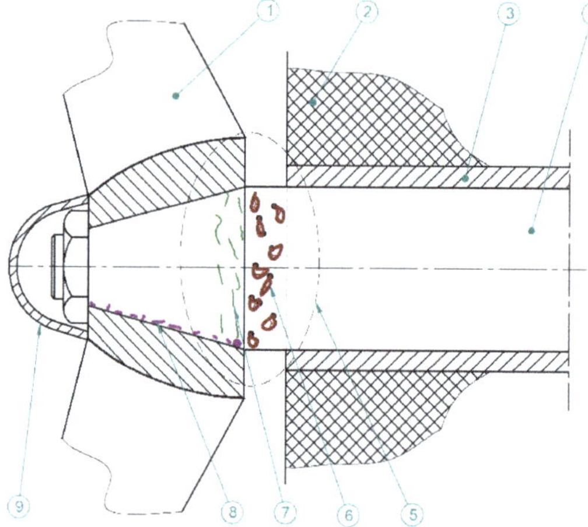

Figure 1. Scheme of screw shaft joint with ship propeller: 1—ship propeller; 2—hull of a ship; 3—bearing sleeve; 4—screw shaft; 5—area of damage; 6—pitting corrosion; 7—fatigue cracks; 8—fretting corrosion; 9—cover.

The propeller shaft, apart from the external forces, is loaded at strengths resulting from the connection joint of the shaft and the propeller. These forces overlap with the external loads that additionally contribute to the development of the abovementioned phenomena, accelerating the process of destruction of the propeller shaft. A scheme of basic forces and moments loading the ship's propeller shaft is presented in Figure 2.

Considering the detailed effects of forces, moments, torques, and vibration occurring in the propulsion system, it may be noticed that the propeller shaft during operation is subjected to a complex state of forces and loads [4,5]. The main load is primarily as a result of the shaft torque, which is created during the power transmission to the propeller. In addition, there are compressive forces as a result of propeller operation that makes the ship move, bending movements as a result of gravity force of the propeller, vibrations arising from a drive system operation, and others. In addition, except external forces, there are internal forces as a result of assembling the propeller and shaft. These forces create stresses in the propeller shaft, which may be the cause of creating surface flaws in the seawater environment (scheme in Figure 1). Increased resistance to the above phenomena can be obtained by surface treatment of the shaft. The literature [6–8] indicates that the burnishing process (described in Section 1.3) enhances the surface resistance on the different external factors.

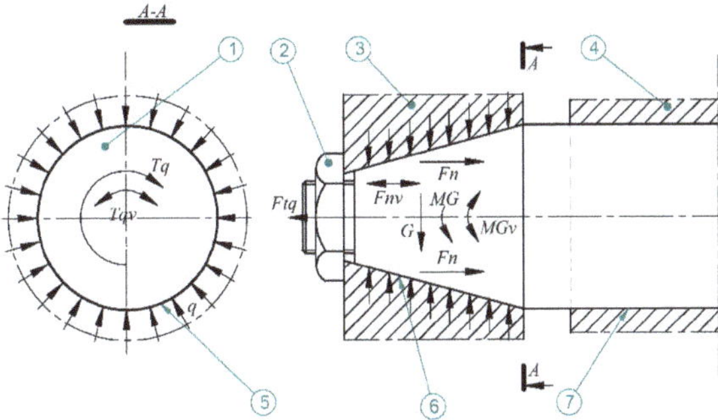

Figure 2. Scheme of forces in screw shaft joint with ship propeller: 1—screw shaft; 2—propeller tightening nut; 3—ship propeller; 4—bearing sleeve; 5—interference surface; 6—conical pressure surface; 7—bearing sliding surface; *G*—gravity force of propeller; *MG*—bending moment from propeller gravity force; *MGv*—bending moment vibration cause by propeller rotation; *Fn*—propelling force; *Fnv*—longitudinal vibrations of the shaft; *Tq*—shaft torque; *Tqv*—vibration of shaft torque; *q*—unit pressure in junction of shaft and propeller hub; *Ftq*—force of assembly tension.

1.2. Bibliography Review

Articles regarding the burnishing of a ship propeller shaft are rarely found in the literature. More articles can be found on other components used in the seawater environment, e.g., pump shafts [9]. The first articles on the ship propeller shafts were written in the 1960s; however, in the following years, this topic has not been addressed. This may be due to fact that this process had been used by the military and the data of this subject have not been available.

The articles [10] analyze the aspects of strengthening and smoothing the shaft surface. In the research for strengthening the process of burnishing, there is the use of geometrically different rollers, which are used for smoothing. This divides the burnishing process into strengthening and smoothing, which is related to the depth of the hardened layer. There are large discrepancies as to the optimal depth of the reinforced layer. Kudryavtsev's study [11] suggests that this layer should be in a range δ = ~0.05r, where r—radius of the shaft.

An additional problem arose before burnishing tools were discussed in the literature [2,10,12]. This wave under the burnishing rollers during the process was being pulled. In shop floor practice, this wave may be easily pulled under burnishing tools and cause destruction in the processed surface, which brings a problem in the shaft manufacturing process. In the literature [2,10], this process is known as the "jumping wave".

In the literature [2,10] for steals with a carbon content in the range of 0.2–0.45% as a result of burnishing, an increase in hardness in the surface layer was obtained by about 1.6 times compared to the hardness after annealing. It is stated that the depth of the burnished layer is not proportional to the obtained hardness, but depends on the diameter of the treated shaft as well as on the geometry of the burnishing element and in particular, on the curvature of the burnishing elements surface. These parameters, together with the burnishing force, create the contact form (surface) on which the depth of hardening depends.

According the literature data [2,10], a cure depth of up to 16% shaft radius may be obtained by the burnishing process. It is necessary to interpret such data carefully because the process of deep hardening of the propeller shaft may cause flaking of the material on its surface. This type of flaw is found in the literature and it is recommended that the shaft surface should be turned or grinded until the removal of this defect and then, the burnishing process is repeated.

In the analyzed articles [13,14], the value of burnishing forces was determined by the Hertz method. This method of burnishing forces determination is not accurate because it is difficult to find the contact area between the tool and the shaft and it is mainly used for single pass machining.

There is very little detail in the literature on the corrosion resistance of burnished surfaces. The available research [7–12] shows that the corrosion resistance of the burnished surface depends on the degree of the cold rolling and, similarly to the fatigue strength, it begins to decrease after exceeding the optimal value. As the corrosion resistance also depends on the type of material and the operating environment, it is difficult to predict its value.

Outside the field of marine technologies, the literature on the fatigue strength of burnished elements is not very large. The research is conducted for various techniques [15–19] (burnishing, shot peening) and materials (stainless steel, nonferrous materials). The comparative analyses of the low plastic burnishing with other technologies in terms of fatigue strength can be found in the literature [16,18]. The results presented in this work show that the burnishing process increases the fatigue strength the most, while there is no information about the limitations concerning low plasticity burnishing. In this research, an increase in fatigue strength to 52% was achieved [17]. In addition, the research is carried out on other processing modifying the surface layer [20] (e.g., laser peening) to improve the properties of manufactured parts.

Recent publications [3,21] indicate that the problem of fatigue strength of marine vehicle propeller shafts is still valid. It is known in the literature [12] that the burnishing treatment improves the surface quality and increases the fatigue strength of the manufactured part, while too high cold rolling range causes the surface to flake and deteriorate its strength. However, there are no precise data on the optimal degree of cold rolling level in relation to the maximum fatigue strength. In addition, the load on the part and the environment in which a given element is operated significantly affects its fatigue strength, while there are few such studies in the literature.

The flaking of the surface after the burnishing process disqualifies the usefulness of the workpiece. On the other hand, the burnishing process increases the fatigue strength only to a certain degree of material cold rolling. The optimum cold rolling value (which maximally increases the fatigue strength) is much lower than the value when flaking occurs and also depends on the operational stresses which will be present in the part. There is a lack of details in the literature on the possible range of the burnishing parameters which could increase the fatigue strength of the workpiece. The lack of this information limits the use of the burnishing process because then it is easy to exceed the optimal degree of cold rolling, which results in lower fatigue strength as compared with its expected value. The aim of the research is to present the possibilities of increasing the fatigue strength of components working in seawater by surface burnishing, also presenting the limitations of this process. As the problem of fatigue strength of marine vehicle propeller shafts is still valid and described in the literature [3,21], the results of the presented research may be useful in designing components for marine vehicles or other applications in the area of ocean technology. The purpose of the article is to test the fatigue strength in seawater of the burnished samples. The results can be useful in designing components for marine vehicles and other applications in ocean technology.

1.3. Introduction to Burnishing Process

The low plasticity burnishing process consists of plastic deformation of the processed surface [12,22,23]. A scheme of this process is in Figure 3. The processed part during plastic deformation is shaped by the tools, which can be rolling or slipping elements, whereby these types of processes are called rolling or sliding burnishing. In rolling burnishing, as working elements roller balls, barrels, and rollers, et al., are used. These elements are usually made from very hard material such as hardened tool steel, sintering carbide, technical ceramic, et al., whereas sliding burnishing tools are usually made from diamond (not using to ferrous materials), regular boron nitride, or other very hard materials. The working surfaces of these tools are usually shaped in the form of a sphere or another similar configuration (for example, paraboloid). The plastic deformation during the burnishing process causes a reduction in the roughness of the surface and moreover, the surface layer of the burnished element also has been strengthened. During this process, the structure of the material changes in the surface layer. The grains of the material are deformed and displaced along the surface to form a surface layer of the different mechanical proprieties in relation to the core, such as hardness, strengthening, et al. Moreover, cold working introduces additional compressive stress into the surface layer.

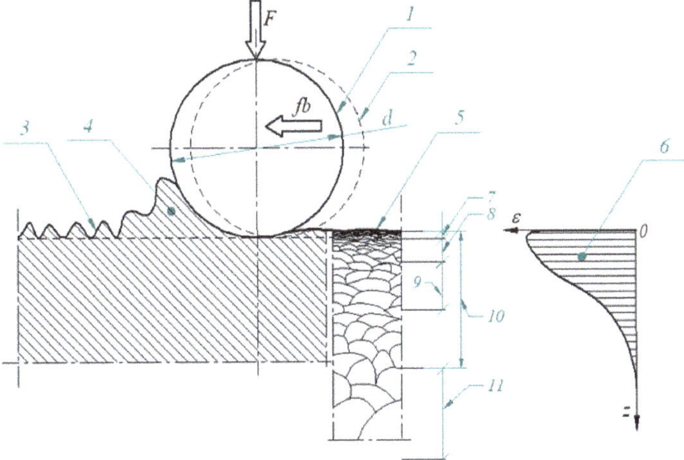

Figure 3. Scheme of low plasticity burnishing process: 1—burnishing ball or roll; 2—previous position of burnishing ball; 3—surface before burnishing; 4—wave of material before burnishing element; 5—surface after burnishing; 6—graph of material strain in parallel to the surface direction; 7—roughness zone; 8—grain fragmentation zone; 9—zone of plastic deformation; 10—zone of elastic deformation; 11-core; F—burnishing force; *fb*—feed of burnishing.

The zones of the surface layer diversified in terms of the properties and structures are shown in Figure 3. The roughness zone includes the range of irregularities of the surface created as a result of processing. The cold plastic deformation process changes the structure of irregularities, mainly by reducing their shape and height. The next zone in the literature [22] is characterized as the grain fragmentation zone, where the material grains were crushed and moved along surface. This zone characterized high hardness and high compressive stresses. Planning the technological process, it should be taken into account that too much deformation degrees and displacement of the material may cause the formation of microcracks in this zone, which may subsequently cause flaking. The graph of strain of the material during the burnishing process in the surface layer is schematic and shown in item 6 in Figure 3.

The next zone is a plastic deformation zone. In this zone, the grains are not crumbled but only plastic deformed and an elongated shape is obtained. In this zone, slight material strengthening occurs and increased hardness and also compressive stress take place. Machining parameters, including the burnishing force, determine the depth of changes in the surface layer of the workpiece and also influence the thickness of particular zones. In the elastic deformation zone, the material is only elastic deformed as a result of the stresses there.

The surface is processed in this way, usually characterized by increased resistance to fatigue loads as well as increased corrosion resistance. The cold rolling degree determined by the burnishing force cannot be too high because it causes surface cracking and next, degreasing of flakes on the burnished surface. This phenomenon in the initial stage is invisible, while damaging the surface and causing a decrease in quality of the processed parts because the fatigue strength corrosion and resistance rapidly go down.

Another phenomenon occurring during the burnishing of soft material (up to 40 Rockwell scale) is the wave of material forming in front of the burnishing tool and moving with it (item 4 in Figure 3). The size of this wave increases with the distance of the tool path. It happens that too large a wave is pulled under the burnishing elements; this causes destruction of the burnished surface. This phenomenon is known in the literature [2,10] as a jumping wave. Usually, counteracting this phenomenon consists of using an additional cutting tool that reduces the excessive size of wave creation.

2. Fatigue Strength Research Test Parameters

The purpose of the experimental tests was to check how the fatigue strength of uniform and joint components operating in the seawater environment are affected by a machining surface process such as burnishing. The cylindrical uniform samples and the conical samples were designed specially to test. Internal stresses of the conical samples were designed by the press sleeve onto the shaft. This press joint was realized using two materials (copper alloy and steel). The idea of the research samples is shown in Figure 4.

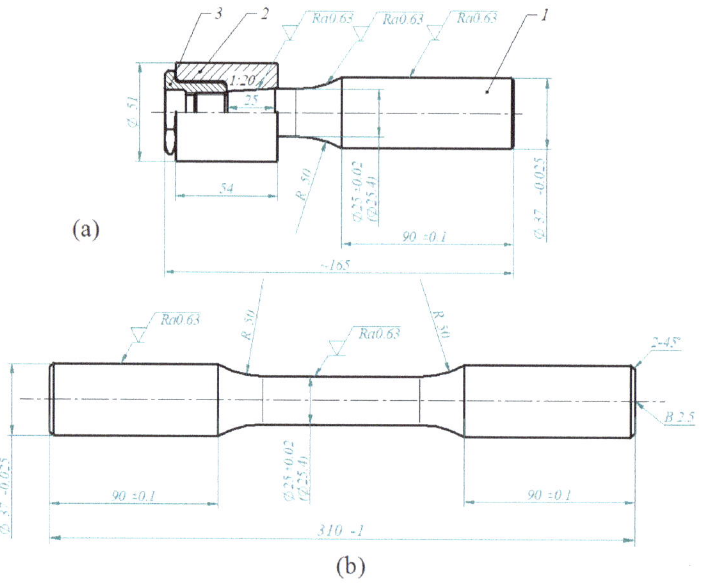

Figure 4. Drawing of test samples (dimensions are specified in [mm]): (**a**) conical sample with assembly joint; (**b**) cylindrical sample; 1—test sample (steel); 2—pressure sleeve (copper alloy); 3—tightening screw.

The tested samples had been prepared as conically and cylindrically shaped with a diameter of 25 mm and were made from steel C35 (1.0501). Steel C35 has been selected for testing because according to the requirements for propeller shaft materials, the steel should have a strength Rm ≥ 430 MPa and be suitable for forging. The chemical composition and mechanical properties in normalized conditions of steel C35 are in Table 1.

The sleeves of the pressing joints of samples were made of a copper alloy. The trade name of this alloy is "Novoston". The chemical composition and mechanical properties of this alloy are presented in Table 1.

Test samples were made by the turning process and then, by surface processing by burnishing or grinding. During the machining, the required dimensional accuracy was to ensure that the deviation of areas of the sample cross-section was not more than 0.5 mm². The samples were rolled, burnishing with variable force value. The range of force changes was from 1.5 to 6 kN. Detailed parameters of the samples processing with the burnishing force are presented in Tables 2 and 3 The feed of burnishing for all samples amounted to 0.214 mm/rev. Burnishing was carried by the three rollers simultaneously, wherein the diameter of rollers was 60 mm. The rounding radius of the burnishing roller profile was R = 20 mm. After the turning process, the surface roughness of the testing shaft determined by the Ra parameter was in a range Ra = 8.77 ÷ 9.07 μm. The sample surface roughness after burnishing according the Ra parameter was in a range Ra = 0.2 ÷ 0.63 μm; the measurement parameters and profile characteristics are presented in Figure 5.

Table 1. Chemical composition and mechanical properties of the sample material.

Chemical composition [%] and mechanical properties in normalized condition of C35 (steel 1.0501)							
C	Si	Mn	P max.	S max.	Cr max.	Ni max.	Cu
0.32 ÷ 40	0.17 ÷ 0.37	0.50 ÷ 0.80	0.040	0.040	0.25	0.25	0.25
Yield Strength Re min		Tensile Strength Rm min		Elongation A5 min		Z min	
314 [MPa]		530 [MPa]		20 [%]		45 [%]	

Chemical composition [%] and mechanical properties in normalized condition for the alloy (75Cu-3Fe-8Al-2Ni-12Mn) ASTM C95700.					
Cu	Mn	Al	Fe	Ni	Si
71	11.0–14.0	7.0–8.5	2.0–4.0	1.5–3.0	0.1
Yield Strength [MPa]		Tensile Strength [MPa]		Elongation [%]	
275		620		15	

Table 2. Results of the fatigue test.

Sample No.	Burnishing Force F [kN]	Sample Diameter d [mm]	Cross-Sectional Area [mm^2]	Stresses of Sample [MPa]	Number of Cycle ×10^6	Condition of the Sample after Testing	Setting Load of the Machine [N]
1		25.00	490.87	292.7	3.69	1	584.5
16		25.00	490.87	259.0	13.5	0	484.4
37		25.00	490.87	222.7	12.0	0	383.4
38	3.0	25.01	491.27	222.7	14.0	1	383.4
46		25.01	491.27	297.2	0.867	1	586.4
47		25.00	490.87	259.9	20.8	0	486.4
48		25.01	491.27	259.0	8.5	1	484.4
3		25.00	490.87	254.1	11.8	0	469.7
21		25.01	491.27	254.1	8.2	1	469.7
23		25.01	491.27	285.1	4.0	1	557.0
2	3.0	25.01	491.27	222.7	25.8	0	383.4
13		25.00	490.87	254.1	11.4	0	469.7
18		25.01	491.27	285.1	8.2	1	557.0
17		25.00	490.87	222.7	8.2	1	383.4
4		25.01	491.27	239.3	11.2	0	433.5
15		25.00	490.87	255.0	9.1	1	475.6
25		25.00	490.87	255.0	10.4	0	475.6
26	6.0	25.00	490.87	255.0	21.3	0	475.6
35		25.00	490.87	255.0	9.2	1	475.6
45		25.01	491.27	270.2	6.8	1	516.8
21		25.01	491.27	270.2	5.9	1	516.8
6		25.00	490.87	239.3	8.9	1	433.5
19		25.00	490.87	239.3	10.3	0	433.5
17		25.01	491.27	266.8	8.5	1	509.9
32	6.0	25.00	490.87	262.9	20.1	0	498.2
35		25.00	490.87	262.9	14.6	0	498.2
14		25.00	490.87	283.5	3.2	1	557.0
20		25.00	490.87	283.5	2.8	1	557.0

0—The sample did not crack during the test; 1—The sample cracked during the test.

Table 3. Calculation results of the fatigue test on the conical samples.

The Kind of Manufacturing Methods	The Stress Range [MPa]	No of Sample		Considered Occurrence					
		Did Not Destroy	Destroyed	Sample Did Not Destroy			Sample Destroyed		
				The Level of Stress	$i \cdot n_i$	$i^2 \cdot n_i$	The Level of Stress	$i \cdot n_{iz}$	$i^2 \cdot n_{iz}$
The grinded conical samples	189.6	2	0	0	0	0	0	0	0
	224.3	1	2	1	1	1	1	2	2
	259	0	2	2	0	0	2	4	8
		$\sum n_i = 3$	$\sum n_{iz} = 4$		$\sum i n_i = 1$	$\sum i^2 n_i = 1$		$\sum i n_{iz} = 6$	$\sum i^2 n_{iz} = 10$
The conical samples burnished with the force F = 3.0 kN (on the entire length l)	222.7	1	1	0	0	0	0	0	0
	259.9	2	1	1	2	2	1	1	1
	297.1	0	2	2	0	0	2	4	8
		$\sum n_i = 3$	$\sum n_{iz} = 4$		$\sum i n_i = 2$	$\sum i^2 n_i = 2$		$\sum i n_{iz} = 5$	$\sum i^2 n_{iz} = 9$
The conical samples burnished with the force F = 3.0 kN (on the $\frac{1}{2}$ length l)	222.7	1	1	0	0	0	0	0	0
	254.1	2	1	1	2	2	1	1	1
	285.1	0	2	2	0	0	2	4	8
		$\sum n_i = 3$	$\sum n_{iz} = 4$		$\sum i n_i = 2$	$\sum i^2 n_i = 2$		$\sum i n_{iz} = 5$	$\sum i^2 n_{iz} = 9$
The conical samples burnished with the force F = 6.0 kN (on the $\frac{1}{2}$ length l)	239.3	1	1	0	0	0	0	0	0
	266.8	2	1	1	2	2	1	1	1
	283.5	0	2	2	0	0	2	4	8
		$\sum n_i = 3$	$\sum n_{iz} = 4$		$\sum i n_i = 2$	$\sum i^2 n_i = 2$		$\sum i n_{iz} = 5$	$\sum i^2 n_{iz} = 9$
The conical samples burnished with the force F = 6.0 kN (on the $\frac{1}{2}$ length l)	239.3	1	0	0	0	0	0	0	0
	255.0	2	2	1	2	2	1	2	2
	270.2	0	2	2	0	0	2	4	8
		$\sum n_i = 3$	$\sum n_{iz} = 4$		$\sum i n_i = 2$	$\sum i^2 n_i = 2$		$\sum i n_{iz} = 6$	$\sum i^2 n_{iz} = 10$

The **bold** values were adopted for the calculation of the estimated fatigue strength.

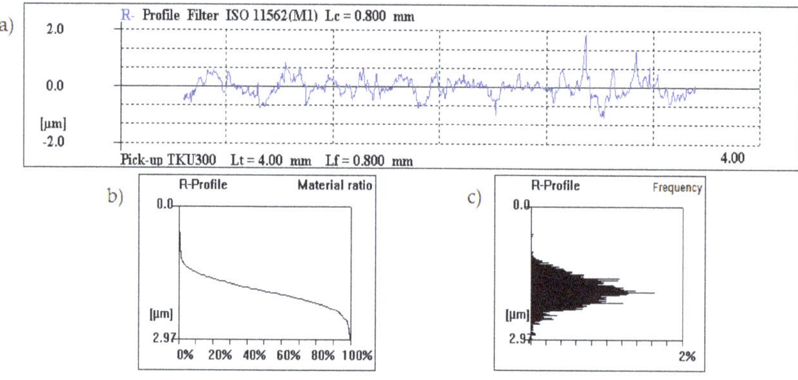

Figure 5. Surface profile of burnishing samples: (**a**) roughness profile; (**b**) material ratio curve; (**c**) frequency density curve.

As a burnishing tool cooling liquid and lubricant during the burnishing process, the machine oil type L-AN 46 (ISO 3448) was used. Figure 6 presents a photograph of the testing samples.

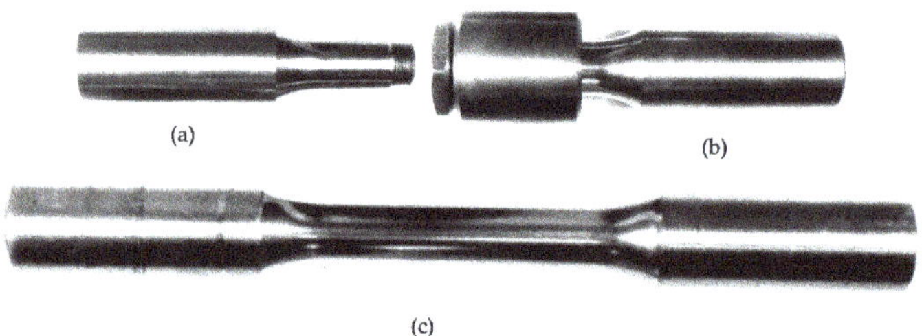

Figure 6. Photographs of samples: (**a**) conical test sample; (**b**) test sample in assembling state as a tightening joint; (**c**) cylindrical sample.

In the conical research samples, additional squeezed stresses were induced by a special sleeve and a nut. The sleeve had been pressed in with the force F = 14.29 kN. The preset amount of interference between the sleeve and sample was performed by the specially designed nut, which had been tightened with the determined torque. The torque value was about M = 60 Nm. In this way, assembled samples were put in the special implement. This implement enabled testing of the samples in the fatigue equipment. The scheme of the implement is presented in Figure 7.

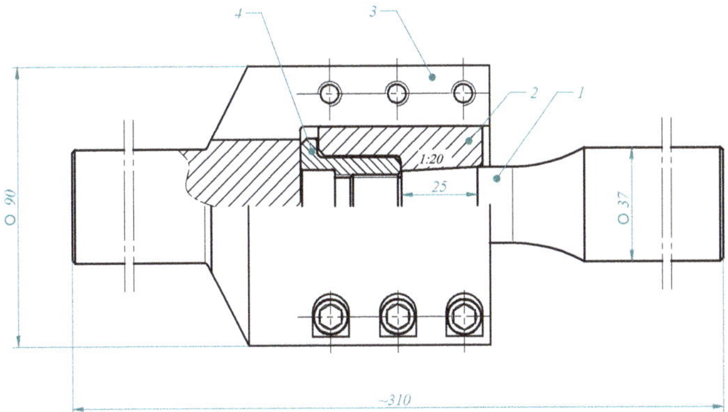

Figure 7. Test sample in the connector for fatigue tests (dimensions are specified in [mm]): 1—test sample; 2—pressure sleeve; 3—connector for fatigue test; 4—tightening screw.

The sample with the sleeve was settled in the special equipment (3). Next, the sample was clamped by tightening by the screws. This prepared sample was settled in the testing stand in the place marked (4) in Figure 8. Then, the sample was placed in a special cover, which enabled fatigue testing in the seawater environment.

Figure 8. Photograph of testing stand for fatigue strength: 1—support bearing; 2—joints for creating testing loads; 3—plates for assembling water cover sheet for seawater; 4—testing sample in the connector; 5—rubber cover sheet for seawater (in open position); 6—clutch; 7—motor; 8—state base; 9—sample before testing; 10—sample after testing.

Seawater was supplied to the cover by a pump from the tank. Figure 9 presents a photograph of the testing stand with the tank of seawater and the system of bending stresses set for the tested samples. In this case, the bending stresses were set by changing the number of weights (6) on the lever (Figure 9).

Figure 9. Fatigue testing stand with equipment for testing in seawater: 1—stand body; 2—weight rod; 3—connecting pipe; 4—water pump; 5—seawater container; 6—weights.

A summary of the conditions carried out for the fatigue research in the testing stand, which is presented in Figures 8 and 9, for the pressed in conical joint and cylindrical samples, is as follows:

- Testing materials: tempered steel C35 (shaft), "Novoston" alloy (sleeve).
- Type of load: rotary bending, $\sigma_m = 0$. R = −1; 1.
- Load frequency: 48.2 cycles/s.
- Adopted base of fatigue cycles: 10^7 cycles.
- Cooling and environment liquid: seawater.
- The cone inclination of shaft–sleeve joint: 1:20 (details of shaft and sleeve dimensions are presented in Figures 4–7).

The research was conducted mainly to determine the permanent fatigue strength.

3. Results and Discussion

The results of resistance to rotational bending are presented in Table 2. The staircase method was applied in the research, wherein the number of destroyed samples was approximately equal to the number of samples not destroyed. The basic criterion of the sample during the fatigue test was occurrence of a sample crack, wherein the fracture should show characteristic features of material fatigue. Using these criteria, the results obtained from the measurements were recalculated and sorted. The results prepared in this way are shown in Table 3.

The results of the tested cylindrical samples in the converted form and sorted are presented in Table 4.

The stair step method requires the number of damaged samples to be equal to the number of samples which have not been destroyed for the same cycle's quantity. In the calculation for the permanent fatigue strength, the number of tests (events) was used, of which frequency of occurrence was smaller. The estimating value of permanent fatigue strength was determined by the following formula [24]:

$$\bar{\sigma} = Z_{gj} = \sigma_0 + \left(\frac{\sum_{i=0}^{q} i n_i}{\sum_{i=0}^{q} n_i} \pm \frac{1}{2} \right) \Delta\sigma \tag{1}$$

where σ_0—the value of the lowest stress level for a rarer occurring test (event); i—number of stress levels; n_i—the frequency of rarer occurring events; $\Delta\sigma$—the difference of the stress level value; "+"—for the damaged samples; "-"—for the not damaged samples.

The estimate average value of standard deviation [24]:

$$S_{(\sigma)} = 1.62 \left[\frac{n \sum i^2 n_i - (\sum n_i)^2}{n^2} + 0.029 \right] \Delta\sigma \tag{2}$$

where σ_0—the value of the lowest stress level for a rarer occurring test (event); i—number of stress levels; n_i—the frequency of rarer occurring events; $\Delta\sigma$—the difference of the stress level value; "+"—for the damaged samples; "-"—for the not damaged samples.

Table 4. Calculation results of the fatigue test on the conical samples.

Type of Technology	Stress Range [Mpa]	Number of Samples		Considered Occurrence					
		Did Not Destroy n_i	Destroyed n_{id}	Samples Did Not Destroy			Samples Destroyed		
				The Level of Stress	$i \cdot n_i$	$i^2 \cdot n_i$	The Level of Stress	$i \cdot n_{id}$	$i^2 \cdot n_{id}$
Ground samples	223	5	0	0	0	0	0	0	0
	229	2	1	1	2	2	1	1	1
	235	1	2	2	2	4	2	4	8
	241	0	3	3	0	0	3	9	27
		$\sum n_i = 8$	$\sum n_{id} = 6$		$\sum i \cdot n_i = 4$	$\sum i^2 \cdot n_i = 6$		$\sum i \cdot n_{id} = 14$	$\sum i^2 \cdot n_{id} = 36$
Samples burnished with force F = 1.5 kN	**263**	2	0	0	0	0	0	0	0
	271	2	1	1	2	2	1	1	1
	279	1	2	2	2	4	2	4	8
	287	0	3	3	0	0	3	9	27
		$\sum n_i = 5$	$\sum n_{id} = 6$		$\sum i \cdot n_i = 4$	$\sum i^2 \cdot n_i = 6$		$\sum i \cdot n_{id} = 14$	$\sum i^2 \cdot n_{id} = 36$
Samples burnished with force F = 2.0 kN	267	3	0	0	0	0	0	0	0
	275	1	1	1	1	1	1	1	1
	283	2	1	2	4	8	2	2	4
	291	0	3	3	0	0	3	9	27
		$\sum n_i = 6$	$\sum n_{id} = 5$		$\sum i \cdot n_i = 5$	$\sum i^2 \cdot n_i = 9$		$\sum i \cdot n_{id} = 12$	$\sum i^2 \cdot n_{id} = 32$
Samples burnished with force F = 2.5 kN	286	4	0	0	0	0	0	0	0
	292	1	2	1	1	1	1	2	2
	298	2	1	2	4	8	2	2	4
	304	0	2	3	0	0	3	6	18
		$\sum n_i = 7$	$\sum n_{id} = 5$		$\sum i \cdot n_i = 5$	$\sum i^2 \cdot n_i = 9$		$\sum i \cdot n_{id} = 10$	$\sum i^2 \cdot n_{id} = 24$
Samples burnished with force F=3.0 kN	292	2	0	0	0	0	0	0	0
	299	2	1	1	2	2	1	1	1
	306	3	2	2	6	12	2	4	8
	313	0	3	3	0	0	3	9	27
		$\sum n_i = 7$	$\sum n_{id} = 6$		$\sum i \cdot n_i = 8$	$\sum i^2 \cdot n_i = 14$		$\sum i \cdot n_{id} = 14$	$\sum i^2 \cdot n_{id} = 36$
Samples burnished with force F=6.0 kN	**274**	2	0	0	0	0	0	0	0
	283	2	1	1	2	2	1	1	1
	292	1	2	2	2	4	2	4	8
	301	0	3	3	0	0	3	9	27
		$\sum n_i = 5$	$\sum n_{id} = 6$		$\sum i \cdot n_i = 4$	$\sum i^2 \cdot n_i = 6$		$\sum i \cdot n_{id} = 14$	$\sum i^2 \cdot n_{id} = 36$

The **bold** values were adopted for the calculation of the estimated fatigue strength.

In Figure 10, the calculated fatigue strength results of the cylindrical samples are presented. The calculation was made on the basis of the data in Table 4 using Equations (1) and (2). The estimated standard deviation has been superimposed on the graph points ($\pm\sigma$). The graph shows that the surface burnishing of the samples significantly increased their fatigue strength compared to the grinding. The increase in fatigue strength of the burnished samples compared to the grinded samples is: for burnishing force 1.5 kN—15%; for burnishing force 2 kN—20%; for burnishing force 3 kN—30%.

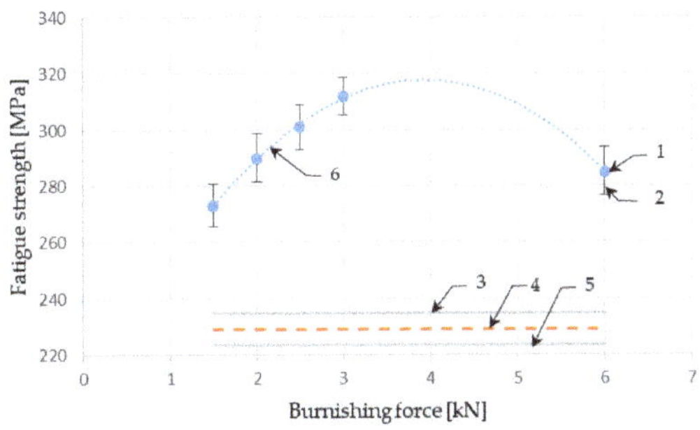

Figure 10. The graph of fatigue strength for the cylindrical samples made with steel (C35) for different burnishing forces and to compare grinding samples. 1—the burnishing; 2—standard deviation of burnishing; 3 and 5—standard deviation of grinding; 4—the grinding; 6—approximating curve as a 2nd degree polynomial.

However, the fatigue strength of samples burnished with the force 6 kN are lower and its value is similar to the samples burnished with the force of 2 kN. It can be assumed that the degree of cold rolling of the material on the surface of the samples increases their fatigue strength only to some extent. Exceeding this critical cold rolling value causes the fatigue strength of the samples to decrease. Excessive force of burnishing causes the creation of microcracks and flakes on the surface. These phenomena cause the formation of notches on the surface, decreasing its fatigue strength. Stress-increased causes enlarge deeper hardening of the surface. The increase in forces enhances the hardening layer but excessive forces cause the formation of cracks on the burnishing surface, which is the reason for the decrease in fatigue strength. Based on the data in Tables 3 and 4, converted according to Equation (1), the comparative results of the fatigue strength for conical and cylindrical samples, which were grinded and burnished with different burnishing forces, are presented in Figure 11. The values of the fatigue results are shown in this figure in the form of points.

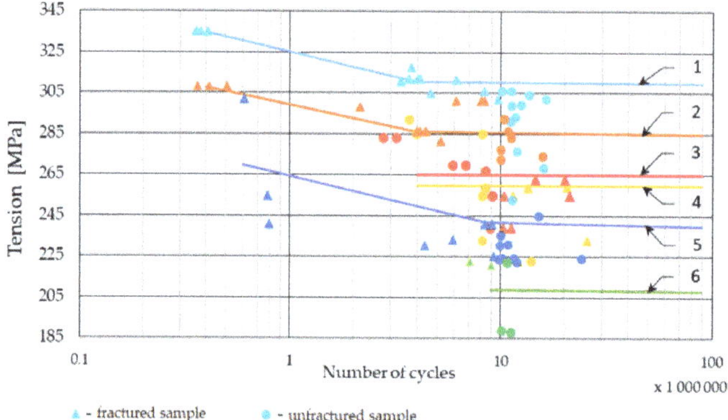

Figure 11. The calculated S–N curves for the samples made with steel (C35): 1—burnishing cylindrical samples by force 3 kN; 2—burnishing cylindrical samples by force 6 kN; 3—burnished conical samples by force 6 kN with a pressed-in sleeve; 4—burnished conical sample by force 3 kN with a pressed-in sleeve; 5—grinded cylindrical samples; 6—grinded conical sample with a pressed-in sleeve.

The burnishing process increases the fatigue strength of the conical samples. Samples burnishing with a pressing force of 3 kN show fatigue strength greater by about 20% compared to the same samples whose surface was grinded. In Figure 11, it can be seen that conical specimens with a pressed-in sleeve have a lower fatigue strength than cylindrical specimens with the same circular cross-sectional area. This applies to both types of surface processing, i.e., grinding and burnishing. The fatigue strength of the cylindrical specimen is about 20% higher than that of the conical sample processed in this same way. It can be assumed that the reduced strength of the conical specimen is due to the additional stress introduced by the pressed-in sleeve. The increase in tension in the sample as a result of overlapping the internal tensions on the external loads had caused the decrease in fatigue strength. As in the case of cylindrical samples, increasing the degree of cold rolling by increasing the crushing force over the extent level does not improve the fatigue strength because samples crushed with 3 and 6 kN do not show significant differences.

It can be assumed that the internal stress caused by the pressed-in sleeve superimposed on the external load, as a result of which the conical sample shows lower resistance to fatigue load. During design works, the occurrence of this phenomenon should be taken into account. When considering the causes of reduced fatigue strength of pressed-in sleeve samples, it is important to take into account that in this joint are the different materials; it may cause, in the seawater, these materials to create a local electric cell, thereby electrochemical processes taking place there may have an effect on the fatigue strength of the joint.

The fatigue fracture of the tested specimen is presented in Figure 12. The area of fatiguing fracture and the immediate fracture (marked 3) can be seen. Comparing these areas may demonstrate that this sample had a large reserve of endurance. There are no visible flaws on the surface on the presented specimen that could be a source of fatigue scrap propagation.

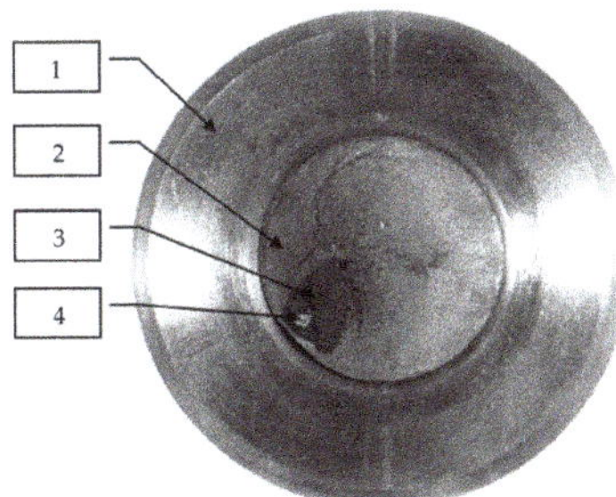

Figure 12. Fatigue fracture for conical sample burnished with the force 3 kN: 1—pressure sleeve; 2—test sample; 3—brittle fracture; 4—material flaw.

Figure 13 presents the fatigue fracture of the tested sample, which was burnished by the force 6 kN. In this figure, surface cracks may be noticed. The cracks appear mainly on the surface and penetrate inside the material. Such a structure of cracks may result from an excessive material of surface layer deformation during the burnishing process. A too high degree of cold working causes microcracks and flaking on the surface; these cracks increase during fatigue load weakening of the sample. These cracks should be investigated in more detail because they greatly reduce the quality parameters of this shaft. Comparing Figures 12 and 13, it can be seen that the surface cracks have occurred only if the surface deformation caused by the burnishing force is too high, which causes a higher degree of cold work.

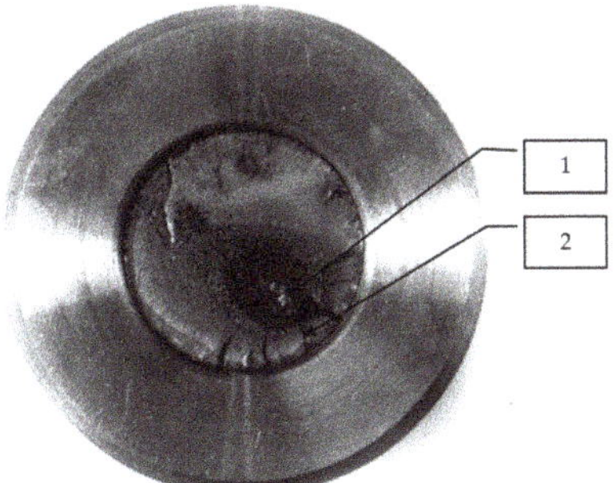

Figure 13. Fatigue fracture for conical sample burnished with the force 6 kN: 1—brittle fracture; 2—cracks on the sample surface.

The fatigue fracture of the conical samples occurred in the area of the larger diameter of the cone near the edge of the pressed-in sleeve. This case confirms occurrence of a stress notch in this place. The impact of the stress notch on the fatigue strength of such components can be limited by strengthening this area by using appropriate design or technological solutions.

4. Conclusion

The article presents research on the fatigue strength of cylindrical and conical samples whose surfaces were burnished with different forces. The samples were tested in an environment of seawater and conical samples were additionally loaded by internal stresses resulting from the pressing in of the testing sleeve. On this basis, these tests may be presented with the following conclusions:

- The sample surface burnishing increases the fatigue strength. The tests showed 30% increase in the fatigue strength of the cylindrical burnished samples in relation to the value of the grinded samples. The fatigue strength increases with the degree of surface cold rolling but only up to a certain value of this cold rolling. Continual increase in the cold rolling degree causes the fatigue strength of the samples to deteriorate, e.g., increasing the burnishing force from 3 to 6 kN, reducing the fatigue strength of the cylindrical samples by 8.7% in relation to the value of the burnished sample with force 3 kN.

- The pressed-in joint has a lower fatigue strength than a uniform sample. The fatigue strength of the cylindrical samples is about 20% greater than the conical specimen with the sleeve pressed in, notwithstanding both specimens had been processed in this same way. This may be due to the additional stress created by the pressed-in sleeve. The surface burnishing of the specimens in such a joint also increases their fatigue strength; the values of fatigue strength of the samples are 20% higher than those in which the surface was grinded. In this case, too much cold rolling on the sample surface does not increase the fatigue strength. This may be due to the formation of microcracks on the surface. The cracks on the surface, decreasing in fatigue strength, of the sample had been observed on the fatigue scrap.

Further research work will be focused on the investigation into the surface layer characteristics being constituted in burnishing processes, realized under differentiated machining conditions.

5. Patent

Method of Shaft Manufacturing

Manufacturing technology of the shafts by burnishing in production conditions may be difficult because burnishing by single tools creates lot of problems such as: balancing of processing forces, clamping and supporting of shafts, and centering on the lathe. In this case, invent equipment was designed, which can turn and burnish simultaneously, and creates an internal balanced layout of processing forces. This equipment enables high quality machining because machining forces do not cause deformation of the workpiece. This equipment can be controlled by a computer system which allows the automation of the process while maintaining very precise machining. The head for integrated machining of stepped shafts is shown in Figure 14. This equipment allows simultaneous machining with cutting and burnishing as well as separate process operating of turning or burnishing. This head should be mounted in the tool holder in such a way that the axis of symmetry of the head should coincide with the rotation axis of the spindle (for the controller, it is usually the Z-axis). The machined shaft is placed inside the head ring. The setting up of the shaft is based on the centers or the lathe chuck and center. The head may be mounted as a floating element and in this case, the head is self-centering relative to the shaft. The control wires of the stepper motor or servomotor and measuring probe should be connected to the machine tool controller. The machining head performs machining simultaneously with three tools, which ensures a balance of cutting forces within the machining place. This type of processing force distribution causes the creation of only torsional stresses in the workpiece, while the shaft does not bend and forces and deformations which could cause shape defects of the workpiece do not occur.

This force distribution is important during the machining of non-rigid shafts because in this case, the unbalanced processing forces deform the machining shaft, which create form errors of the manufactured part. The move out cylinder (6) system in burnishing tools enables independent operation of cutting and burnishing tools. This system enables planning of the burnishing process only on selected surfaces and the turning or burnishing process may be carried out in many passes. The measuring probe (8) placed on the head verifies the obtained dimensional effects of machining and in case of deviations, may create applications of appropriate corrections. The above head covered by the patent application [25] increases the productivity of stepped shafts manufacturing with high slenderness by eliminating many technological operations and human labor from the currently used technologies.

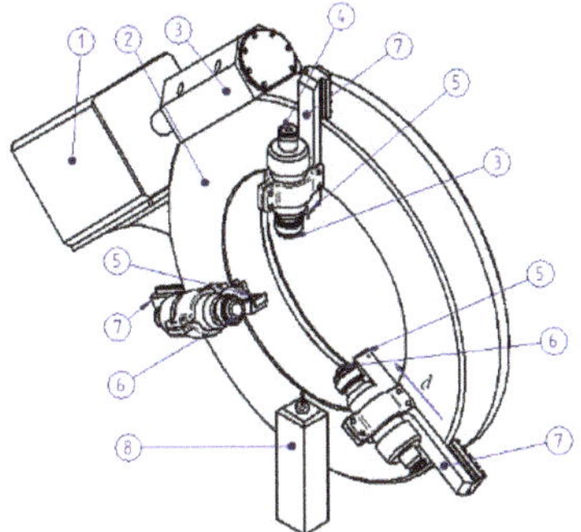

Figure 14. Scheme of the equipment for machining stepped shafts on computerized numerical controlled machine tools by turning and burnishing: 1—stepper motor; 2—equipment body; 3—worm gear; 4—quick-release joint of the hydraulic system; 5—cutting insets; 6—burnishing balls; 7—slide-way of tools; 8—measuring probe; d—direction of tool offset.

Figure 15 presents an example instruction sheet of the processing of the marine propeller shaft. The process of shaft manufacturing is planned with the use of special equipment shown in Figure 14. The whole process of turning and burnishing may be carried out in a computerized numerically controlled machine tool. The fine surface burnishing process enables manufacturing of the shaft in one turning and burnishing operation, eliminating the grinding operation as a finishing. The measuring probe and machining control system allows the realization of the machining process, which has been executed with the accuracy required for a marine propeller shaft.

T04	Process engineer	Instruction Sheet 2a	Part name: Tall shaft							Oper. No. 010		Part Number
			Operatin name: Turning and Burnishing							Sheet 6a		8557/411-1-5
Operation course			D: B	L	i	V [m/min]	n [rpm]	f [mm/rev]	F [kN]			
1. Mount the tool equipment according to the instruction sheet for machine setup.										Machine type		Station
2. Monut the shaft on the machine tool, secure with counter supports.										TCC-161		MP 131
3. Turn and burnish the shaft cylidrical diameter ⌀770 and cone according to the schematic diagram.			⌀770	3560	1	29	12	0.35	145			
4. Remove the counter supports, move the support together witch the tool equimpment for burnishing ⌀775 remount the counter supports and check it.												
5. Burnish a diameter of ⌀775 according to the schematic diagram.			⌀775	840	1	29	12	0.35	145			
6. Check the quality of performed operations. **Note**: If the required dimensions are not achieved, repeat the burnishing process with the reduced force by 50%.												

Workshop auxiliaries

Oper.	Name	Symbol	Amount
	HEAD	UNT	1
	Lathe chuck	PULFS800 /105-SD	1
	Fixed center	PZKa A6	1
	Rotary center	PZKk6	1
	Counter support		2
	Cone convergence gauge	MS016	1

Dimension tolerances IT9

Figure 15. Exemplary instruction sheet for the operation of strengthening burnishing of a B557 type marine tail shaft with the use of the equipment for turning and burnishing.

Author Contributions: Conceptualization, S.D., W.P. and B.Ś.; methodology, S.D., W.P. and B.Ś.; software, S.D., W.P. and B.Ś.; validation, S.D., W.P. and B.Ś.; formal analysis, S.D., W.P. and B.Ś.; investigation, S.D., W.P. and B.Ś.; resources, S.D., W.P. and B.Ś.; data curation, S.D., W.P. and B.Ś.; writing—original draft preparation, S.D., W.P. and B.Ś.; writing—review and editing, S.D., W.P. and B.Ś.; visualization, S.D., W.P. and B.Ś. All authors have read and agreed to the published version of the manuscript.

Funding: This research received no external funding.

Conflicts of Interest: The authors declare no conflict of interest.

References

1.   Carion, G.; Simihiati, B. *Damage to and Repairs of Propeller Shaft and Liners*; Bulletin technique du Bureau Veritas: Levallois-Perret, France, 1985.
2.   Shook, L.L.; Long, C.I. Surface Cold rolling of Marine Propeller Shafting. 1957. Available online: https://www.shotpeener.com/library/pdf/1957004.pdf (accessed on 16 October 2020).
3.   Kettrakul, P.; Promdirek, P. Failure analysis of propeller shaft used in the propulsion system of a fishing boat. *Mater. Today* **2018**, *5*, 9624–9629. [CrossRef]
4.   Huang, Q.W.; Liu, H.G.; Ding, Z. Dynamical response of the shaft-bearing system of marine propeller shaft with velocity-dependent friction. *Ocean Eng.* **2019**, *189*, 106399. [CrossRef]
5.   Yao, H.Y.; Cao, L.L.; Wu, D.Z.; Yu, F.X.; Huang, B. Generation and distribution of turbulence-induced forces on a propeller. *Ocean Eng.* **2020**, *206*, 107255. [CrossRef]
6.   Karpenko, G.V.; Pogoretskii, R.G.; Sirak, Y.M. The influence of surface hardening by burnishing on the fatigue resistance of shafts with press fits in a corrosive medium. In *The Protection of Metals from Corrosion-Mechanical Damages*; Profizdat: Moscow, Russia, 1970; pp. 113–115.
7.   Ryabchenkov, A.V. *Corrosion Fatigue Strength*; Mashgnitz: Moscow, Russia, 1953; [in Russian].
8.   Vaingarten, A.M.; Goman, G.M.; Mendelson, I.V. The corrosion fatigue strength and corrosion resistance of 35 steel specimens in relation to the hardening method. *Tfkh. Sudostroyeinie.* **1973**, *3*, 54–57.
9.   Dyl, T.; Dolny, K. The influence of the Burnishing on Technological Quality of Elements of Part Shipping Machines. *J. KONES* **2010**, *17*, 89–95.
10. Toshio, K.; Eiichi, S.; Bunzai, A. On Fatigue Test of Materials with Cold Rolling for the purpose of Strengthening of Marine Propeller Shaft. In *Japan Shipbuilding and Marine Engineering*; Japan Association for Technical Information: Tokyo, Japan, 1966.
11. Kudryavtsev, I.V. Selection of the basic parameters for strengthening shafts by burnishing, I.V. Kudryavtsev. *Vestnik Mashinostroenia* **1983**, *4*, 8–10, [in Russian].
12. Przybylski, W. *Low Plastic Burnishing Processes, Fundamentals, Tools and Machine Tool*; Institute for Sustainable Technologies-National Research Institute: Radom, Poland, 2019.
13. Klocke, F.; Lierman, J. Roller burnishing of hard turned surfaces. *Int. J. Mach. Tools Manuf.* **1998**, *38*, 419–423. [CrossRef]
14. Skalski, K.; Morawski, A.; Przybylski, W. Analysis of contact elastic-plastic strains during the the process of burnishing. *Int. J. Mech. Sci.* **1995**, *37*, 461–472. [CrossRef]
15. Travieso-Rodríguez, J.A.; Jerez-Mesa, R.; Gómez-Gras, G.; Llumà-Fuentes, J.; Casadesús-Farràs, O.; Madueño-Guerrero, M. Hardening effect and fatigue behavior enhancement through ball burnishing on AISI 1038. *J. Mater. Res. Technol.* **2019**, *8*, 5639–5646. [CrossRef]
16. Avilés, A.; Avilés, R.; Albizuri, J.; Pallarés-Santasmartas, L.; Rodríguez, A. Effect of shot-peening and low-plasticity burnishing on the high-cycle fatigue strength of DIN 34CrNiMo6 alloy steel. *Int. J. Fatigue* **2019**, *119*, 338–354. [CrossRef]
17. Yuan, X.L.; Li, C.Y. An engineering high cycle fatigue strength prediction model for low plasticity burnished samples. *Int. J. Fatigue* **2017**, *103*, 318–326. [CrossRef]
18. Aviles, R.; Albizuri, J.; Rotriguez, A.; Lopezde Lacalle, L.N. Influence of low-plasticy burnishing on the high-cycle fatigue strength of medium carbon AISI1045 steel. *Int. J. Fatigue* **2013**, *55*, 230–244. [CrossRef]
19. Zielecki, W.; Bucior, M.; Trzepieciński, T.; Ochał, K. Effect of slide burnishing of shoulder fillets on the fatigue strength of X19NiCrMo4 steel shafts. *Int. J. Adv. Manuf. Technol.* **2020**, *106*, 2583–2593. [CrossRef]
20. Karthik, D.; Kalainathan, S.; Swaroop, S. Surface modification of 17-4 PH stainless steel by laser peening without protective coating process. *Surf. Coat. Technol.* **2015**, *278*, 138–145. [CrossRef]
21. Arisoy, C.F.; Başman, G.; Kelami, Ş.M. Failure of a 17-4 PH stainless steel sailboat propeller shaft. *Eng. Fail. Anal.* **2003**, *10*, 711–717. [CrossRef]
22. Przybylski, W.; Dzionk, S. Hibrid processing by turning and burnishing of machine components. In *Advances in Manufacturing*; Hamrol, A., Ciszak, O., Legutko, S., Jurczyk, M., Eds.; Springer International Publishing: Cham, Switzerland, 2018; pp. 587–598.

23. Ścibiorski, B.; Dzionk, S. The roughness of the hardened steel surface created by rolling burnishing process. *Solid State Phenom.* **2015**, *220*, 881–886. [CrossRef]
24. Dyląg, Z.; Orłoś, Z. *Fatigue Strength of Materials*; Wydawnictwa Naukowo-techniczne: Warszawa, Poland, 1962; [in Polish].
25. Dzionk, S.; Przybylski, W. Way and Device for Single Operation Turning or Burnishing of High Slender Shafts on a Numerically Controlled Machine Tool. PL. Patent 221524 B1, 22 May 2019.

Publisher's Note: MDPI stays neutral with regard to jurisdictional claims in published maps and institutional affiliations.

Article

Tool Wear Prediction in Single-Sided Lapping Process

Norbert Piotrowski

Department of Manufacturing and Production Engineering, Faculty of Mechanical Engineering,
Gdańsk University of Technology, 80233 Gdańsk, Poland; norbert.piotrowski@pg.edu.pl

Received: 24 August 2020; Accepted: 23 September 2020; Published: 25 September 2020

Abstract: Single-sided lapping is one of the most effective planarization technologies. The process has relatively complex kinematics and it is determined by a number of inputs parameters. It has been noted that prediction of the tool wear during the process is critical for product quality control. To determine the profile wear of the lapping plate, a computer model which simulates abrasive grains trajectories was developed in MATLAB. Moreover, a data-driven technique was investigated to indicate the relationship between the tool wear uniformity and lapping parameters such as the position of conditioning rings and rotational speed of the lapping plate and conditioning rings.

Keywords: abrasive machining; single-sided lapping; tool wear; machine learning

---

1. Introduction

Lapping is the basic planarization process which allows achieving a high degree of flatness and parallelism of machined workpieces. It has become a crucial finishing technology in manufacturing several parts, e.g., valve plates, ceramic sealing rings or sliced silicon wafers. The lapping process is very complex and is influenced by many technologically-based conditions. The model of the process consists of several components: lapping plate, abrasive slurry, workpieces and machine tool. These elements decisively affect the mechanism of surface formation and determine workpieces quality, tool wear and overall process efficiency. In addition, the input factors of lapping process can be classified as controllable and uncontrollable. The first group includes mainly machining parameters, i.e., lapping pressure, rotational speed of the lapping plate and conditioning rings, types of abrasive grains, tool condition and machining time. The uncontrollable parameters include the ambient temperature, grains size distribution, vibrations, internal stresses, etc. [1].

In lapping, the lapping plate has a significant impact on the dimensional and shape accuracy, as well as on the surface quality of the workpieces. The material, design and technological parameters of the lapping plates are selected depending on the desired material removal rate, required surface finish, expected flatness and hardness and geometry of the lapped workpieces [2,3]. The most important features of the tool are the mechanical properties, type of material, structure, macrogeometry and topography of the active surface [2,4]. The mechanical properties and structure of the lapping plate material have a decisive influence on the abrasion resistance and quality of the machined surfaces. The macrogeometry of the lapping plate surface has an impact on the distribution of the abrasive slurry in the working gaps, and thus on the topography of the lapped surface. Equally significant is the hardness parameter. Using a lapping plate with too high hardness causes most abrasive grains to roll, resulting in stress-induced microcracks and sedimentation of the grains in workpieces. However, this also leads to lower tool wear. On the other hand, a plate made of a material with lower hardness keeps the abrasive grain on the surface and causes more sliding movements, and the loss of material results from scratching [5,6].

In addition, the flatness of the workpieces and the proper surface finish are maintained while maintaining appropriate conditions of the process and frequently correcting the lapping plate flatness. Excessive wear of the tool by the workpieces causes the following flatness errors: concavity, convexity or the axial runout of the active surface [7]. Therefore, it is important to properly check and adjust the flatness of the lapping plate before and during the machining. Additionally, the machine operator should periodically lap the plate with fine abrasive grains to obtain a surface free from larger scratches.

The main subject of this article is the analysis of single-sided lapping systems in the aspect of wear uniformity of the lapping plate. The remainder of this paper is structured as follows. Section 2 reviews the related work on tool wear prediction in planarization technologies. Section 3 describes in detail the lapping mechanism and the influence of process parameters on the accuracy of machining. A kinematics-based model of the lapping plate profile wear, which allows analyzing lapping system parameters, as well as to control the shape of the tool in terms of the required flatness, was developed. Section 4 introduces the data driven approach to predictive modeling of lapping plate wear. Section 5 closes with conclusions and future work.

2. Related Work

Previous studies on different planarization technologies, such as lapping, face grinding with lapping kinematics or chemical-mechanical polishing (CMP), have focused primarily on analysis of abrasives path distribution, determining the technological aspects of machining new materials or study on thermal influence on the process efficiency and the workpieces quality [8–10]. Many studies have been published on mechanisms of material removal and the effects of input variables on material removal rate (MRR). Moreover, today's progress in the field of machine learning allows classifying predictive methods for estimating MRR into two categories: physics-based and data-driven methods. Both types of models can also be used to determine the rate of the lapping plate wear.

2.1. Physics-Based Models

Chang et al. [11] developed the mechanical material removal mechanisms using concepts of two-body vs. three-body abrasion and ductile vs. brittle machining position information. MRR was captured in the model through the change in abrasive size distribution. Kling et al. [12] described the material removal mechanisms of the workpieces and the lapping plate in single- and double-sided lapping. Presented examples of calculations showed how to control the shape of the plate by controlling the workpiece distribution and the speed ratio. Nguyen et al. [13] studied the pad wear non-uniformity by combining the kinematic motions and contact time. It was indicated that the cutting path density and the contact time near the center of pad are higher than near the edge of the pad. Barylski and Deja [14] presented a model which based on the kinematics of lapping and local shape errors of the tool estimates the profile error. Simulation results have been compared with experimental results obtained during machining of silicon wafers. Lee et al. [15] introduced a model that estimates the MRR using a modified Preston equation. Three parameters, namely relative velocity distribution, normal contact stress distribution and chemical reaction rate distribution, were considered for obtaining the MRR profile in the copper CMP process.

2.2. Data-Driven Models

Wang et al. [16] developed a data-driven technique based on the deep belief network to reveal the relationship between MRR and polishing parameters such as pressure and rotational speeds of the wafer and pad. To predict MRR in CMP, Li et al. [17] developed a decision tree-based ensemble learning algorithm. Three decision tree-based machine learning algorithms were combined using two stacking techniques. Yu et al. [18] introduced a novel physics-informed machine learning approach. They combined a physics-based model with a data-driven model of MRR in CMP. The machine learning algorithm predicts the asperity radius and asperity density of the polishing pad is estimated by the physics-based model.

3. Physics-Based Model

The basic kinematic parameters that the machine operator can control during single-sided lapping are the rotational speeds of the lapping plate n_t and the radial position R of the condition rings (together with separators and workpieces). In more complex lapping machines with forced drive, it is also possible to independently set the rotational speed of the condition rings n_s. The analysis in this section aims to determine the effect of these parameters on the uniformity of the lapping plate wear.

3.1. Analysis of the Angular Velocities of the Conditioning Ring and the Lapping Plate

Based on the developed kinematic equations [19], trajectories cut on the lapping surface by abrasive grains, which are fixed in the conditioning ring and workpieces, can be generated. To analyze the effect of velocities of the lapping plate ω_t and the conditioning ring ω_s, the dimensionless parameter k_1 is determined:

$$k_1 = \frac{\omega_s}{\omega_t} \tag{1}$$

Changing the parameter k_1 allows obtaining a wide range of trajectories. If the k_1 parameter is positive, it means that the lapping plate and the conditioning ring rotate in the same direction, while, when the k_1 parameter is negative, the directions of their rotations are opposite. In addition, a detailed analysis of generated trajectories allowed drawing the following conclusions:

- For $k_1 < 0$, generated trajectories are epicycloids, initially stretched and then interlaced.
- For $k_1 > 1$, generated trajectories are hypocycloids, initially stretched and then interlaced.
- For $0 < k_1 < 1$, generated trajectories are pericycloids.

Moreover, it was noticed that, if the parameter k_1 is a rational number, the trajectories become closed curves and the time of one complete cycle depends on the velocities ω_s and ω_t.

Figures 1–4 show the influence of the parameter k_1 on the shape of the trajectory, the density of trajectories on the lapping plate showed in the form of a color map and the distribution of relative velocities in time $v(t)$. The simulations were performed for the conditioning ring distant from the center of the lapping plate by $R = 117$ mm and for one abrasive grain distant from the center of the guide ring by $r = 50$ mm. Since in each of the analyzed examples the parameter k_1 is a rational number, it was assumed that the simulation time t is equal to the time of one complete cycle, when generated trajectories are closed curves, i.e., $T = 60$ s. It can be observed that, with changing the speed ratio of the conditioning ring ω_s and the lapping plate ω_t, and thus with the change of the shape of generated cycloid, the trajectory density on the plate changes as a function of the radius of the lapping plate R_d. For epicycloid trajectories (Figures 1 and 2), the highest density occurs near the outer diameter of the lapping plate. The hypocycloid trajectories (Figure 3) cause a higher density of trajectories closer to the inner diameter of the plate. The most uniform trajectories distribution occurs in the case of pericycloids (Figure 4), i.e., when the parameter k_1 is positive and lower than 1. On the basis of the relative velocities in time distribution $v(t)$, the ratio of the minimum and maximum velocities v_{min}/v_{max} can be determined. It can be also noticed that the change of the relative velocity v of any point P is a periodic function. The periodicity of the function is determined from the condition of equal length of the radius $R_p(t)$ and it is the time after which the length of the vector $R_p(t)$ is equal to the initial length ($t = 0$). The duration of one cycle T_c is calculated with Equation (2) and the results for different parameter k_1 are shown in Table 1.

$$T_c = \frac{2\pi}{|\omega_s|} \tag{2}$$

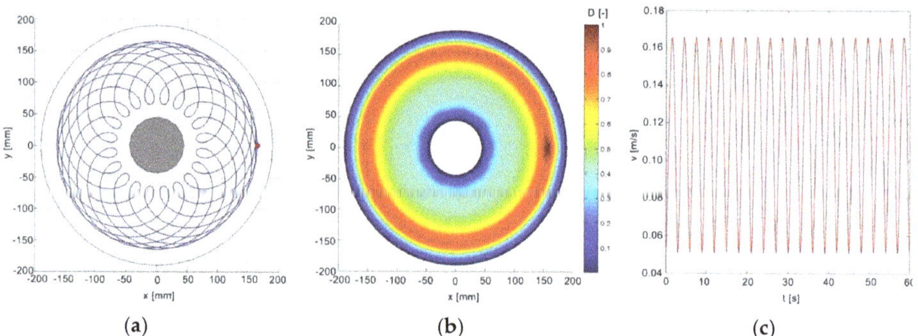

Figure 1. The influence of the parameter $k_1 = -20/9$ on: (**a**) the shape of the abrasive grains trajectory; (**b**) the trajectories density on the lapping plate; and (**c**) the distribution of relative velocities $v(t)$.

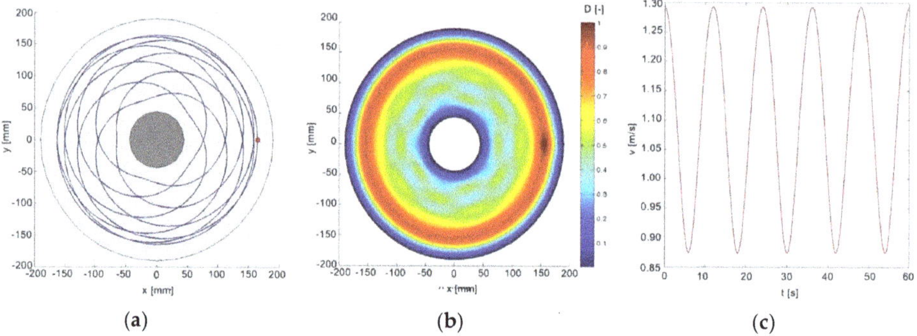

Figure 2. The influence of the parameter $k_1 = -5/9$ on: (**a**) the shape of the abrasive grains trajectory; (**b**) the trajectories density on the lapping plate; and (**c**) the distribution of relative velocities $v(t)$.

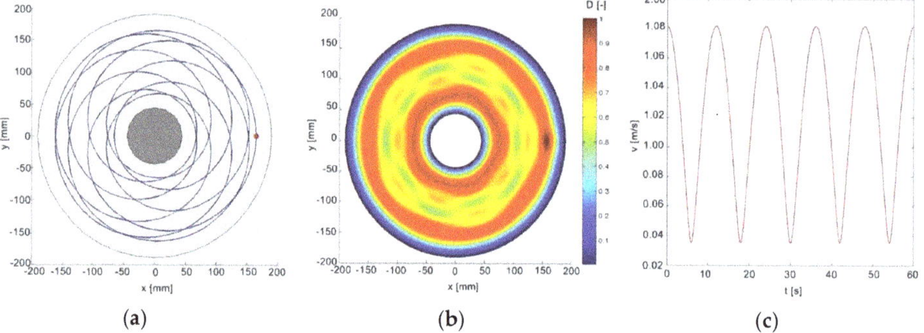

Figure 3. The influence of the parameter $k_1 = 5/9$ on: (**a**) the shape of the abrasive grains trajectory; (**b**) the trajectories density on the lapping plate; and (**c**) the distribution of relative velocities $v(t)$.

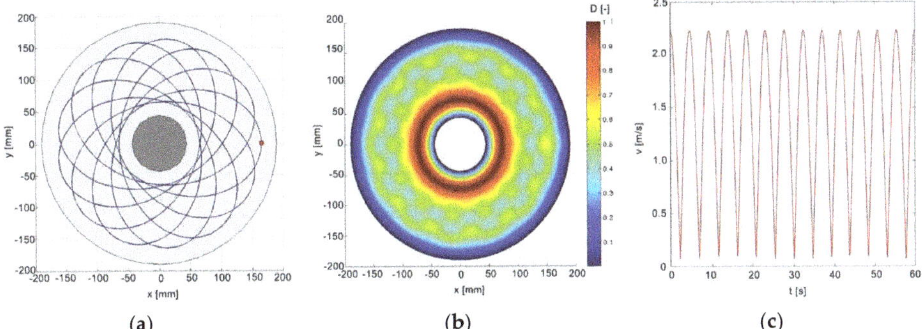

Figure 4. The influence of the parameter $k_1 = 13/9$ on: (**a**) the shape of the abrasive grains trajectory; (**b**) the trajectories density on the lapping plate; and (**c**) the distribution of relative velocities $v(t)$.

Table 1. Cycle duration for different parameters k_1.

k_1	$-20/9$	$-5/9$	$5/9$	$13/9$
T_c (s)	3	12	12	4.62

3.2. Analysis of the Radial Position of Conditioning Rings

The main aim of using conditioning rings in lapping process is to even the active surface of the lapping plate. In standard single-sided lapping system, the rings are driven by the friction torque between the lapping plate and the rings. In more complex machine tools, the rings can be driven separately. The radial position of the conditioning rings is usually changed manually. Shifting the rings, filled with the separators and workpieces, towards the inner or outer diameter of the lapping plate changes the contact intensity in these areas. Moving the conditioning ring towards the outer radius increases velocity in the radial section of the lapping plate. Radial displacement of the ring causes changing the parameter k_1, which is related to the change of the velocity ω_s. Decreasing the parameter ω_s due shifting the conditioning ring to the center of the lapping plate increases the contact intensity in the outer surface of the lapping plate.

Figure 5 shows the distribution of the average relative velocity v_{sr} as a function of the radius R_d for different values of the parameter k_1 and for different radial positions of the conditioning ring R. The simulations were performed during time $t = 60$ s for one point distant from the center of the conditioning ring by $r = 50$ mm. It can be observed that, when the directions of conditioning rings and the lapping plate rotations are opposite, and when the value of the ring speed is correspondingly higher than the value of the lapping plate speed (Figure 5a), the average relative speed v_{sr} decreases towards the outer radius of the lapping plate R_d. Moreover, for parameters k_1 lower than zero (Figure 5a,b), increasing the radial position of the ring R causes an increase in the mean velocity v_{sr}. Conversely, for k_1 greater than zero (Figure 5c,d), increasing the radial position of the ring R causes a decrease in the mean velocity v_{sr}.

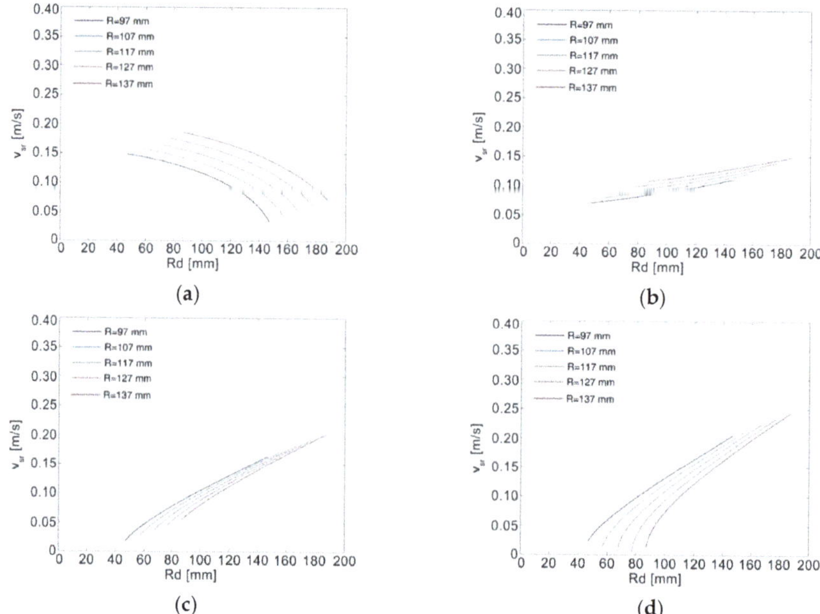

Figure 5. Average velocity distribution v_{sr} as a function of radius R_d for different parameters k_1: (**a**) $k_1 = -20/9$; (**b**) $k_1 = -5/9$; (**c**) $k_1 = 5/9$; and (**d**) $k_1 = 13/9$.

3.3. Tool Wear Uniformity Prediction

The physics-based methods can be classified into two sub-categories: kinematics-based models and kinematics-mechanistic models. In kinematic based models, it is assumed that MRR is the product of cutting by single abrasive grains and the wear intensity depends on a contact intensity of the tool with the workpieces through the lapping abrasive grains. To determine the contact intensity, the trajectories of the randomly grains are calculated using the kinematic equations. The interpolation function is used to change these trajectories into a set of points, which are equally spaced from each other. Then, the lapping plate surface is divided into small squares with the same area. To calculate a trajectories density, a statistics function is used to count the total number of points within each square of the lapping plate surface [19].

The limitation of kinematic-based models is that the material removal mechanisms is not taken into account [18]. This can be complemented by kinematics-mechanistic modeling approaches, which take into account both kinematics and contact mechanics (Figure 6). One of the commonly used methods to predict MRR of the abrasive processes is the tribological model developed by Preston. The model relating relative velocity to pressure is known as Preston's equation [20]:

$$\frac{dH}{dt} = K \cdot p \cdot v \tag{3}$$

It can be assumed that for constant conditions Preston's coefficient K and force per unit area p are constant in time. The relative velocity of lapping v can be calculated from the developed kinematic equations [19]. By combining Preston's model (Equation (3)) with the contact intensity, which depends on the lapping kinematics, the material removal rate by abrasive grains within the contact area A_i of the lapping plate can be determined as:

$$H_i = \frac{K \cdot p \cdot \sum v_i}{A_i} \tag{4}$$

To describe the uniformity of the lapping wear, the parameter U was defined:

$$U = \left(1 - \frac{S_D}{\overline{H}}\right) \cdot 100\% \tag{5}$$

where S_D is the standard deviation of the trajectory density and $\overline{H}$ is the average value of the material removal rate.

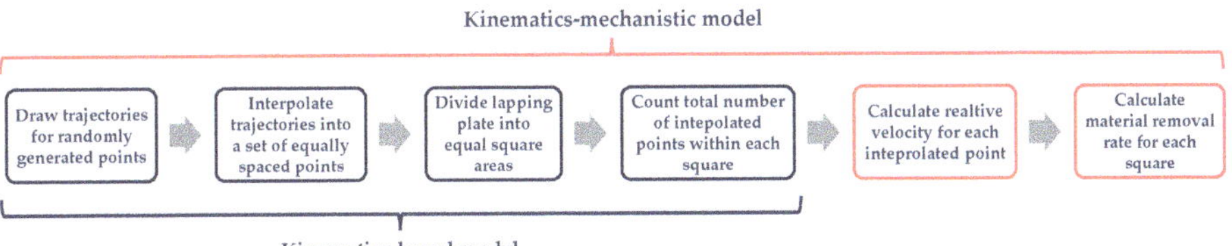

Figure 6. Kinematics and kinematics-mechanistic models comparison.

During the lapping process, there are more than one million active abrasive grains. However, because of the very long time of the calculation, an appropriate grains number, which can reflect the same regularity as the real number, had to be determined. It was observed that, for 1000 randomly distributed particles, uniformity was stable and further increase did not affect the value of parameter U [19]. A MATLAB program was developed to simulate the trajectories, their distribution and to calculate the uniformity parameter. Uniformity was calculated on the entire lapping plate with an internal diameter of 88 mm and an outer diameter of 350 mm. The simulation time was 60 s. The simulation results are shown in Figure 7. It can be observed that, to obtain higher uniformity values in a standard lapping system, rotational speed ratio k_1 should be in range 0.5–0.9. Moreover, the higher uniformity can be obtained for the central position of the conditioning ring on the lapping plate. Figure 7b shows a wear distribution on the lapping plate for the parameters $R = 117$ mm and $k_1 = 7/9$ that could be considered as optimal, i.e., when the parameter U is the highest. A significant decrease in the uniformity ($U < 30\%$) was observed when the conditioning ring was shifted towards the internal radius and the parameter $k_1 > 1.5$ or when the ring was shifted towards the external radius and parameter $k_1 < -1$.

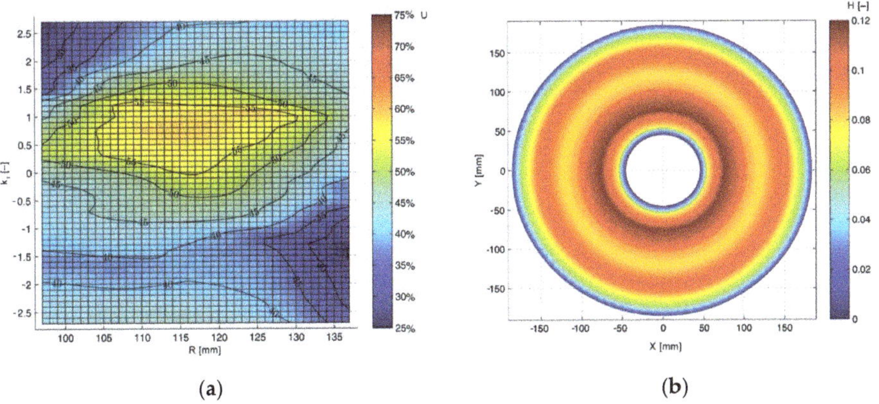

(a)　　　　　　　(b)

Figure 7. Simulation results: (**a**) distribution of the wear uniformity U with relative to the parameter k_1 and the radial position of the conditioning ring R; and (**b**) distribution of the lapping plate wear for parameters $R = 117$ mm and $k_1 = 7/9$.

4. Data-Driven Model

Physics-based models are very effective in predictive MRR, but they involve strong assumptions, which may not hold true for certain conditions. To address these assumptions and limitations, data-driven predictive modeling methods can be used to predict MRR or tool wear. Data-driven predictive modeling methods are built upon statistical methods or machine learning algorithms, which can adaptively improve their performance with each new data sample. While machine learning methods may require large volumes of training data, they can make predictions without predetermined mechanistic relationship and system behaviors.

Experimental datasets in lapping plate wear are commonly limited in size because the tests require many and very precise measurements that are difficult to do directly on the machine tool. In addition, the tool wear over time is low due to the lapping plate hardness. However, this fact does not make machine learning impractical for predictive modeling of tool wear. There are a few common approaches that can help with building predictive models from small datasets. First, the more complex is the model, the more it is prone to overfitting. This can be avoided by cross-validation, regularization, feature selection and bucketing, which aim to reduce complexity and increase bias. With small datasets, it is important to rely on a simple classifier models, e.g., k-Nearest Neighbors (kNN), short decisions trees or Naïve Bayes. In general, these simple models are able to learn less from data than more complicated algorithms, e.g., neural networks, making them less susceptible to overfitting. Another solution for small amounts of data could be using ensemble methods. Ensembles are machine learning methods for combining multiple machine learning algorithms, which are known as base learners, into one learning algorithm. The main aim is to improve the performance of predictive models by reducing: a predictive variance by randomization, a predictive bias by boosting or both by stacking [21,22].

In this study, a data-driven method was demonstrated to predict lapping plate wear uniformity in single-sided lapping process. The data were collected from a typical single-sided lapping process conducted on the machine tool equipped with a standard lapping plate, three conditioning rings and a system that allows controlling the direction and rotational speed of the rings. The dataset contains results of lapping plate profile wear for different conditions, i.e., lapping workpieces made of different materials, different geometries, using different abrasive grains with different concentrations and for various initial state of the tool. Moreover, to determine the profile wear of the lapping plate in experimental tests, the contact method was used. This method is based on determining in the first stage the actual shape of the tool surface. The amount of wear at the measured point is determined by the difference in measurements before and after the lapping process. The measurement method was carried out with the use of a special measuring tool, which is a part of the laboratory stand. The tool consists of five dial gauges, on which the radial distance on the lapping plate can be adjusted. In these experimental tests, an equal distance between the sensors was assumed, which was 35 mm. Measurements were made on each of the 12 segments of the plate and the average value was saved in the computer program. Some of these results have already been published [19].

Due to the small number of results, only two predictors were established in the model. These are the main operation parameters of the single-sided lapping process, i.e., speed ratio k_1 and radial position of the condition rings R. Based on the measurements of the lapping plate profile wear, uniformity U was calculated. Moreover, to simplify the model, the uniformity U was bucketed into five classes, which are presented in Table 2.

Table 2. Different classes of tool wear uniformity U.

Class	A	B	C	D	E
U	55–59%	60–64%	65–69%	70–74%	75–79%

Based on the dataset presented in Figure 8, it can be observed that the uniformity U differs from the results of the kinematic-mechanistic approach. Although the distribution looks similar, the real uniformity of the lapping plate wear is relatively higher. The highest result was close to 80%, while for the simulation the highest uniformity U was around 60%. On the one hand, differences may result from the limitations of the physics-based model. On the other hand, another factor affecting the experimental results is inaccurate measuring method. Measurements in experiments were not taken over the entire length of the diameter, but only at a few points of the lapping plate.

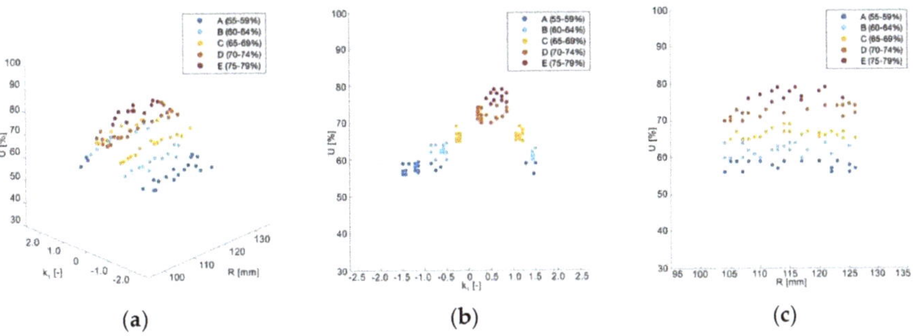

Figure 8. Dataset used in data-driven model: (**a**) overall view; (**b**) influence of the parameter k_1 on the uniformity of U; and (**c**) influence of the parameter R on the uniformity of U.

Moreover, the dataset was randomly divided into two groups including training dataset (80%) and test dataset (20%). Three algorithms with a use of MATLAB Machine Learning Toolbox were tested to predict tool wear uniformity: the k-Nearest Neighbors (when $k = 3$), decision trees and the Naïve Bayes classifiers (optimized with Kernel smoothing). Models were trained and tested 10 times and then the average accuracy was calculated. Decision tree shows the best overall accuracy (94.8%), then Naïve Bayes (89.5%) while kNN induction exhibits the worst one (68.5%). In addition, to investigate the performance of the proposed models, the multi-class confusion matrixes are shown in Figure 9. The confusion matrix counts the total number of observation values of the classification model, which were classified into the right and wrong categories. Coordinate axis of the confusion matrix represents the actual label of classification, and the horizontal axis represents the predicted class.

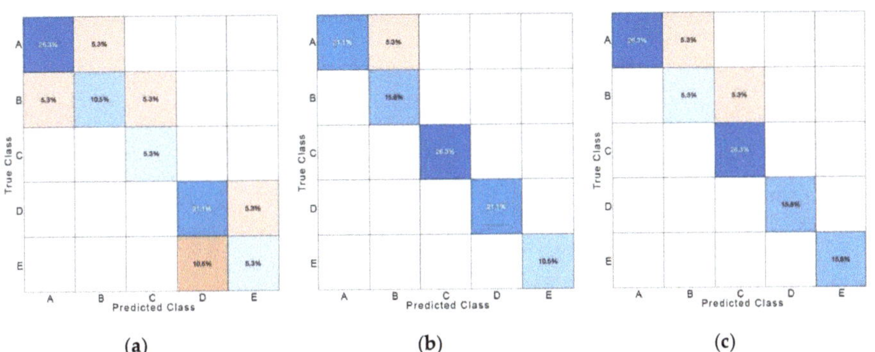

Figure 9. Confusion matrices summarizing the accuracy results for: (**a**) kNN algorithm; (**b**) decision tree model; and (**c**) Naïve Bayes classifier.

5. Conclusions

The analysis of the single-sided lapping kinematics showed that the basic operation parameters influence the relative speed and significantly affect the distribution of abrasive grains trajectories. In practice, the selection of kinematic parameters is made largely intuitively, based on some experience or from the literature and then tested experimentally. The kinematic model developed in MATLAB enables simulations of a single-sided lapping process, generate trajectories of randomly chosen abrasive grains and determines the basic kinematic parameters of the process, such as relative speed v. Moreover, it calculates the uniformity of lapping plate wear U for any shape and size of the workpieces, their arrangement in separators and the assumed number of abrasive grains. Simulations of physics-based model showed that the highest uniformity of lapping plate wear ($U = 58\%$) was obtained for the central position of the conditioning ring and when the ratio of rotational speeds of the ring and the lapping plate equals to $k_1 = 0.75$. Moreover, a data-driven method to predict lapping plate wear uniformity was developed. Simplifying the model by bucketing the uniformity U into five classes allows using supervised machine-learning algorithms for tool wear prediction. The best accuracy was obtained for the decision tree classifier with result of accuracy 94.8%. This model had difficulty with prediction of class A, which results from too narrow range of predictors and experimental studies.

The accuracy of the data-driven method is calculated based on the number of correctly predicted classes out of all predictions. In addition, the accuracy of this method depends largely on the accuracy of the measuring method used in experimental research. Direct comparison of the lapping plate wear results in experimental tests and physics-based model is ineffectual. It should be noted that simulated in physics-based model tool wear was a dimensionless parameter and many input parameters were not taken into account. Moreover, the difference between the simulation and experimental results presented in the article is due to the limitations of the physics-based model and the fact that the wear of the lapping plate in the laboratory tests was measured only at five points.

Due to the complexity of the lapping process and interactions involved, not all the issues related to the topic were resolved. Both developed models have still limitations which, due to the importance of the problem of uneven wear of the lapping plate, should be improved. Owing to wide machine learning possibilities, the data-driven model will be expanded with new predictors: initial state of the lapping plate, lapping time and contact pressure. This will require increasing the training datasets, which requires more experimental research.

Funding: This research received no external funding.

Conflicts of Interest: The author declares no conflict of interest.

List of Abbreviations and Symbols

A	contact area of the lapping plate (m^2)
CMP	chemical-mechanical polishing
H	lapping plate wear removal rate (μm/min)
k	number of nearest neighbors
kNN	k-Nearest Neighbors algorithm
K	Preston's coefficient
k_1	speed ratio of the conditioning ring and the lapping plate
MRR	material removal rate
n_t	rotational speeds of the lapping plate (rpm)
n_s	rotational speeds of the conditioning ring (rpm)
p	contact pressure (MPa)
R	radial position of the condition rings (mm)
R_D	radius of the lapping plate (mm)
S_D	standard deviation of the trajectory density
t	simulation time (s)
T	time of cycle, when generated trajectories are closed curves (s)
T_c	time of one cycle in $v(t)$ function (s)
U	lapping plate wear uniformity (%)
v	relative velocity of the analyzed point P (m/s)
v_{sr}	average relative velocity (m/s)
ω_t	angular velocity of the lapping plate (rad/s)
ω_s	angular velocity of the conditioning ring (rad/s)

References

1. Deaconescu, A.; Deaconescu, T. Experimental and statistical parametric optimisation of surface roughness and machining productivity by lapping. *Trans. FAMENA* **2015**, *39*, 65–78.
2. Uhlmann, E.; Ardelt, T.G. Influence of kinematics on the face grinding process on lapping machines. *CIRP Ann.* **1999**, *48*, 281–284. [CrossRef]
3. Kasai, T.; Horio, K.; Karaki-Doy, T.; Kobayashi, A. Improvement of Conventional Polishing Conditions for Obtaining Super Smooth Surfaces of Glass and Metal Works. *CIRP Ann.* **1990**, *39*, 321–324. [CrossRef]
4. Marinescu, I.D.; Uhlmann, E.; Doi, T. *Handbook of Lapping and Polishing*; Taylor & Francis Publishing House: New York, NY, USA, 2007.
5. Marinescu, I.D.; Rowe, W.B.; Dimitrov, B.; Inasaki, I. *Tribology of Abrasive Machining Processes*; William Andrew Publishing Ltd.: Norwich, UK, 2004.
6. Feld, M.; Barylski, A. Lappen ebener Flachen mit Zweimetall-Scheiben. *Werks. u. Betr.* **1990**, *123*, 933–936.
7. Deja, M. Correlation between shape errors in flat grinding. *J. Vibroeng.* **2012**, *14*, 520–527.
8. Liu, H.K.; Chen, C.A.; Chen, W.C. Diamond lapping of sapphire wafer with addition of graphene in slurry. *Procedia Eng.* **2017**, *184*, 156–162. [CrossRef]
9. Lu, L.Y.; Fang, C.F.; Shen, J.Y.; Lu, J.; Xu, X.P. Analysis of path distribution in lapping and polishing with single fixed abrasive. *Key Eng. Mater.* **2014**, *589*, 475–479. [CrossRef]
10. Deja, M.; List, M.; Lichtschlag, L.; Uhlmann, E. Thermal and technological aspects of double face grinding of Al_2O_3 ceramic materials. *Ceram. Int.* **2019**, *45*, 19489–19495. [CrossRef]
11. Chang, Y.P.; Hashimura, M.M.; Dornfeld, D.A. An investigation of material removal mechanisms in lapping with grain size transition. *J. Manuf. Sci. Eng.* **2000**, *122*, 413–419. [CrossRef]
12. Kling, J.; Matthias, E. Workpiece material removal and lapping wheel wear in plane and plane-parallel lapping. *CIRP Ann.* **1986**, *35*, 219–222. [CrossRef]
13. Nguyen, N.Y.; Zhong, Z.W.; Tian, Y. An analytical investigation of pad wear caused by the conditioner in fixed abrasive chemical–mechanical polishing. *Int. J. Adv. Manuf. Technol.* **2015**, *77*, 897–905. [CrossRef]
14. Barylski, A.; Deja, M. Wear of a tool in double-disk lapping of silicon wafers. In Proceedings of the 5th International Manufacturing Science and Engineering Conference (MSEC 2010), Erie, PA, USA, 12–15 October 2010; pp. 301–307.
15. Lee, H.; Jeong, H. A Wafer-Scale Material Removal Rate Profile Model for Copper Chemical Mechanical Planarization. *Int. J. Mach. Tools Manuf.* **2011**, *51*, 395–403. [CrossRef]
16. Wang, P.; Gao, R.X.; Yan, R. A deep learning-based approach to material removal rate prediction in polishing. *CIRP Ann.* **2017**, *66*, 429–432. [CrossRef]
17. Jia, X.; Di, Y.; Feng, J.; Yang, Q.; Dai, H.; Lee, J. Adaptive Virtual Metrology for Semiconductor Chemical Mechanical Planarization Process Using GMDH-Type Polynomial Neural Networks. *J. Process Control* **2018**, *62*, 44–54. [CrossRef]
18. Yu, T.; Li, Z.; Wu, D. Predictive modeling of material removal rate in chemical mechanical planarization with physics-informed machine learning. *Wear* **2019**, *426*, 1430–1438. [CrossRef]
19. Barylski, A.; Piotrowski, N. Non-conventional approach in single-sided lapping process: Kinematic analysis and parameters optimization. *Int. J. Adv. Manuf. Technol.* **2019**, *100*, 589–598. [CrossRef]
20. Evans, J.; Paul, E.; Dornfeld, D.; Lucca, D.; Byrne, G.; Tricard, M.; Klocke, F.; Dambon, O.; Mullany, B. *Material Removal Mechanisms in Lapping and Polishing*; University of California: Berkeley, CA, USA, 2003; pp. 611–633.
21. Zhou, Z.H. *Ensemble Methods: Foundations and Algorithms*; CRC Press: Boca Raton, FL, USA, 2012.
22. Li, Z.; Wu, D.; Yu, T. Prediction of Material Removal Rate for Chemical Mechanical Planarization Using Decision Tree-Based Ensemble Learning. *J. Manuf. Sci. Eng.* **2019**, *141*, 031003. [CrossRef]

Article

Studying the Effect of Working Conditions on WEDM Machining Performance of Super Alloy Inconel 617

Stefan Dzionk and Mieczysław S. Siemiątkowski *

Department of Manufacturing and Production Engineering, Faculty of Mechanical Engineering,
Gdansk University of Technology, 80-233 Gdańsk, Poland; sdzionk@pg.edu.pl
* Correspondence: msiemiat@pg.edu.pl; Tel.: +48-668-194-998

Received: 19 August 2020; Accepted: 7 September 2020; Published: 9 September 2020

Abstract: Wire electrical discharge machining (WEDM) has been, for many years, a precise and efficient non-conventional manufacturing solution in various industrial applications, mostly involving the use of hard-to-machine materials like, among others, the Inconel super alloys. The focus of the present study is on exploring the effect of selected control parameters, including pulse duration, pulse-off time and the dielectric flow pressure on the WEDM process performance characteristics of Inconel 617 material, such as: volumetric material removal rate (MRR), the dimensional accuracy of cutting (reflected by the kerf width) and surface roughness (SR). The research experiment has been designed and carried out using the response surface methodology (RSM) accordingly with the Box–Behnken design scheme. The results of experiments derived in the form of a fitted regression model have been subjected to the analysis of variance (ANOVA) tests. Thus, the variable process parameters and the relevant interactions between them, characterized by a significant influence on the values of the derived output responses, could be explicitly determined.

Keywords: wire EDM; process parameters; Inconel 617 super alloy; DoE; RSM method; Box–Behnken design; ANOVA; performance measures; material removal rate; surface roughness

---

1. Introduction

Rapid technological development has recently created a demand for new materials that would have properties enabling their use under harsh operating conditions, including the high temperature, aggressive chemical environment, at variable loads and others. Conditions of such a nature occur, inter alia, in gas turbines, nuclear reactors, power generators equipment or chemical industry installations [1]. One such material is an alloy based on nickel and chromium, which also contains other elements such as: niobium, iron, aluminum [2], commercially termed as "Inconel". These materials are characterized by temperature strength, low thermal diffusivity, high hardness and others, to be classified as super alloys [1,2]. These properties cause that this material is used among others in gas turbine disks or responsible aircraft components. The Inconel super alloys belong to the group of hard-to-machine materials, which is considered as one of their major drawbacks. This is due to high content of abrasive particles in this alloy, its low thermal and electrical conductivity, high tendency to welding to the tool, and forming built-up edges, etc. [1,2].

Electrical discharge machining (EDM), and its variety implemented using a wire, and referred to as wire electrical discharge machining (WEDM), are among those types of technologies where the above-mentioned characteristics of machined material have no major impact on the final machining result. This type of non-conventional processing has been known since the 1940s, and for many decades applied in modern industry as CNC controlled precision machining, including non-standard applications, as in e.g., [3], despite the substantial disadvantage of the technology concerning high energy consumption and its quite low efficiency.

There are relatively few reports in the literature on modelling or investigating the efficiency and effectiveness of machining Inconel alloys with EDM methods. One is the work of Hevidy et al. [1], which highlights the development of mathematical models for correlating the relationships of various process parameters in wire EDM machining of Inconel 601 on the characteristics of performance, based on the response surface methodology approach (RSM), approach to the design of experiment (DoE), and its analysis. It studies the influence of four process factors on the volumetric material removal rate (MRR), wear ratio (work related volumetric MRR as a percentage of wire MRR) and surface roughness (SR). The presented results show the variability of the volumetric MRR parameter in the range of 4-8 mm^3/min and the Ra parameter within 0.25–7.5 μm. The authors conclude that the Ra parameter drops down with the increase in wire tension, that would, however, require consideration of whether the value of waviness filtered out the roughness parameter value of Ra. The experimental results in [4] revealed that pulse-on time, pulse-off time and servo voltage greatly affect the MRR and SR characteristics. In the research paper [5] in particular, a successful attempt was made to optimise the above-mentioned process parameters for increased MRR and SR in wire EDM of Inconel 625 alloy, using Taguchi methodical scheme and ANOVA.

The paper [6] reports on the influence of the newly type of copper-based SiC electrode, developed, among others, for wire-cut EDM applications, on material removal efficiency and on surface finish in main and trim cuts. As a result, they obtained an increase in cutting efficiency by an average of 16%, and the Ra parameter value decreased by about 17% on average. In the research paper [7], an impact of chief process parameters of WEDM of Inconel 617 on the micro-hardness of machined surfaces and the density of created surface cracks was investigated. In the conclusion, it is stated that the value of micro-hardness and number of micro-cracks increase with the value of pulse current and pulse duration. The study of Li et al. [2] presents the characteristics of surface integrity versus discharge energy. In this work, surface roughness is shown to be equivalent for parallel and perpendicular wire directions, and its mean value can be significantly reduced through lowering the discharge energy. They compared the surface structure in the main and the trim cuts and reported on observed thick white layers, predominantly discontinuous and non-uniform with confined micro voids, at the increased energy of discharge. An adequate review of the state-of-art research on surface integrity issues considering also WEDM processes of nickel-based super alloys can be found in [8]. In [9], the authors report in turn on performed experimental investigation with the aim to determine the main WEDM parameters, which contribute to recast layer formation (between 5 and 9 μm in average thickness) in Inconel alloy material. It was found that average recast layer thickness increased primarily with energy per spark, peak discharge current, and pulse duration. Selected results concerning the influence of processing conditions on stereometric surface parameters and the morphology of the surface layer in WEDM of hard machinable materials, including in part to Inconel alloys, are provided in [10].

At the same time, numerous reports can be found in the literature which tackle the issues related to machining other type alloys with EDM. Mostafafor and Vahedi [11] in particular presents results of tests on the magnesium alloy in the view of its effective machining. Various aspects of the accuracy and efficiency of the WEDM process of magnesium alloys are addressed in [12]. Likewise, a paper of Markopoulos et al. [13,14] discusses the results of experimental study on selecting process conditions that determine the quality of WEDM of an aluminum alloy. Discussion of issues of choosing the right conditions for WEDM of titanium alloys can be found in [15]. In the paper of Rao and Krishna [16] the optimization capabilities of some performance characteristics in WEDM machining of reinforced ZC63 alloy metal matrix composite (MMC).

The analyzed literature reports do not present a complete description of the machining process characteristics of Inconel alloys, and results are focused on selected input parameters, tested in relatively limited ranges, which is the missing element, particularly within the current knowledge on machining Inconel alloys by wire-EDM. The purpose of this research is to determine the process efficiency, its accuracy and the workpiece surface integrity in relation to three process parameters, i.e., pulse-on time (T-on), pulse off time (T-off) and the dielectric flow pressure (DFP). Those parameters were chosen because, besides the voltage, they are the basic machining parameters having the influence on the machining efficiency. The flushing pressure, in turn, effects the intensity of the removal of WEDM process debris from the machining zone, which inhibits the processing intensity.

2. Materials and Methods

The electrical discharge manufacturing process uses electrical discharges between the workpiece and the working electrode to the removal of material form processed part component. The energy of a single discharge causes a local melting of the material, which then is removed by the flowing dielectric. Such a material removal phenomenon means that processing materials must possess a conductivity higher than 10^{-2} S/cm. In the WEDM method, the electrode is usually made of a brass or the other alloy, and it is moved by the machine heads, which are steered by the CNC controller.

The schematic diagram of the device is shown in Figure 1a. The wire moving in the material cuts the kerf larger than the its diameter; the scheme of this process is shown in Figure 1b. The difference between of the kerf width and the wire diameter is called spark gap, and it depends on the process parameters. Sparkling is caused by the generator, which creates electrical impulses that is shown schematically in Figure 1c. The effectiveness of a single discharge can be determined by an empirical equation [17,18].

$$Q_i = k \cdot E_i \tag{1}$$

where: Q_i—volume of removed material [mm^3], E_i—single discharge energy, k—proportionality factor of the cathode and the anode.

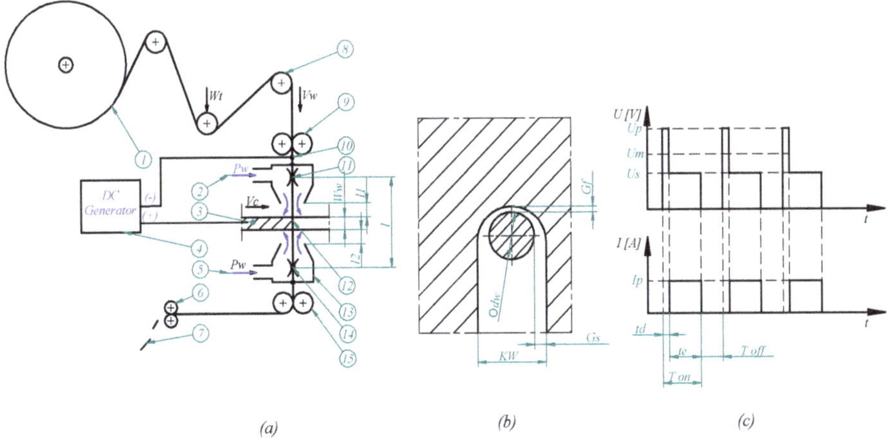

(a)   (b)   (c)

Figure 1. Wire electrical discharge machining components: (**a**) scheme of the process, (**b**) view of the wire in the kerf, (**c**) graph of current impulse run, 1—wire reel, 2—upper inlet of flushing liquid (dielectric), work piece, 4—supply unit, 5—lower inlet of flushing dielectric, 6—wire chopper, 7—cut wire scrap, 8—rolls of wire tensioning unit, 9—upper wire drive, 10—current pick up, 11—upper wire guide 12—wire, 13—the nozzle of lower flushing, 14—lower wire guide, 15—bottom wire drive, Vw—wire feeding, Wt—wire tension, Vc—workpiece feed, Pw—flushing water pressure, Ww—workpiece width, l—distance before guides, l1—upper flushing die gap, l2—lower flushing die gap, dw—wire diameter, Gf—front spark gap, Gs—lateral spark gap, KW—cutting kerf, Up—pulse peak voltage, Um—average working voltage, Us—discharge voltage, Ip—pulse peak current, td—ignition delay time, te—pulse discharge time, T-on—pulse on time, T-off—pulse off time.

Energy of a single discharge can be calculated by the formula:

$$E_i = \int_0^t U(t) \cdot I(t) dt \tag{2}$$

where: E_i—single discharge energy, t—discharge time, $U(t)$—voltage, $I(t)$—current intensity.

As voltage and current are varied with time, they could be replaced with effective values, defined as below:

$$U_{eff} = \sqrt{\frac{1}{t} \int_0^t U^2(t) dt} \tag{3}$$

$$I_{eff} = \sqrt{\frac{1}{t} \int_0^t I^2(t) dt} \tag{4}$$

where: t—impulse time, U_{eff}—effective voltage I_{eff}—effective current, $U(t)$—voltage, $I(t)$—current intensity.

Single impulse energy is by:

$$E_{efi} = U_{eff} \cdot I_{eff} \cdot T_{on} \tag{5}$$

where: E_{efi}—energy of single impulse, T_{on}—impulse time, U_{eff}—effective voltage, I_{eff}—effective current.

Power generated during machining in the working zone, that enables the removal of all material content, can be found by the following equation:

$$P_{eff} = E_{efi} \cdot f = U_{eff} \cdot I_{eff} \cdot T_{on} \cdot \frac{1}{T_{on} + T_{off}} \tag{6}$$

where: P_{eff}—power of discharges [W], E_{efi}—energy of single impulse [J], T_{on}—impulse time [μs], f—frequency [s^{-1}], T_{off}—break between pulses [μs], U_{eff}—effective voltage [V] I_{eff}—effective current [A]. Total amount removing material can be, in turn, expressed as:

$$Q_T = P_{eff} \cdot k \cdot r \tag{7}$$

where: Q_T—total amount removing material [mm^3/s], k—factor of proportionality cathode and anode, r—the coefficient of efficiency of electrical discharge.

Demand for power may be determined, using:

$$P_T = \frac{P_{eff}}{\mu} \tag{8}$$

where: P_T—demand for power, P_{eff}—power of discharges, μ—the discharge generator efficiency.

In this research, the empirical equation has been utilized to determine the MRR levels (see Section 3.1), since the values for the related factors presented in the above given formulas are difficult to be determined [19].

Based on the Equations (6) and (7), the parameters T-on and T-off were, as seen further, selected for the tests, as they have direct impact on the machining performance. The WEDM process tests were performed on an AccuteX AU-300iA (Taichung, Taiwan) machine, and the process parameters were set in accordance with the designed experiment, and described further on.

After starting, the value of the linear feed it was gradually increased until the value at which the process was no longer stable. The maximum linear cutting feed value gained at which the EDM process maintained its stability was recorded, with several repetitions to confirm this required state.

In this experimental study three controllable parameters of the WEDM process, viz., pulse-on time (T-on), pulse-off time (T-off) and dielectric flow pressure (DFP) were considered as input variable parameters (independent variables). The encoding levels and actual values of those process parameters are listed in Table 1.

Table 1. Variable machining parameters and their levels for the wire electrical discharge machining (WEDM) research of Inconel 617.

Parameter	Unit	Low Level (−1)	Mid-Level (0)	High Level (+1)
Pulse-on time T-on	μs	0.6	0.9	1.2
Pulse-on time T-off	μs	6	10	14
Dielectric flow pressure (DFP)	MPa	0.3	0.5	0.7

Other process factors that might affect the performance measures are summarized in Table 2. Consistently, they were held constant throughout the experiments as far as practicable.

Table 2. Constant machining conditions in this research study.

Parameter (Input Constant)	Unit	Value
Cutting mode	–	DC
Voltage	V	50 ± 3
Arc on time	μs	0.8
Arc of time	μs	8
Wire tension	N	15
Wire feed	mm/s	13
Dielectric temperature	°C	22 ± 0.5
Dielectric conductivity (ION+)	μS	10–16
Wire diameter	mm	$0.25_{-0.002}$
Material of wire	–	CuZn37
Tensile strength of wire	N/mm2	900
Conductivity of wire acc. IACS	%	22
Upper flushing die gap	mm	0.4
Lower flushing die gap	mm	0.05

The variations ranges of input parameters were established on the basis of experience and some exploratory tests, considering the capacity of the applied WEDM machine and successful cutting with the avoidance of wire tear and interruption of the process.

An 8 mm thick rolled plate made from Inconel 617 (material certified by BIBUS Metals AG, Switzerland) was used in the tests. Chemical composition and mechanical properties of this material are presented in Table 3.

Table 3. Chemical composition of Inconel 617 alloy and its mechanical properties.

Chemical Composition of Inconel 617 Alloy [WT. %]													
C	Mn	Fe	S	Si	Cu	Ni	Cr	Al	Ti	Co	Mo	P	B
0.09	0.11	1.25	<0.001	0.04	0.03	53.92	21.98	1.12	0.34	11.47	9.65	<0.002	<0.001

Mechanical Properties (in temperature 26 °C)							
Yield Stress	Tensile Strength	Elongation	Reduction of Area	Hardness	Density	Electrical Resistivity	Melting Range
443 [MPa]	827 [MPa]	62 [%]	56 [%]	57.6 [BHN]	8.36 [Mg/m^3]	1.22 [μΩ·m]	1332–1380 [°C]

WEDM tests were made as the cuts in the plate, 15 mm long. A metallographic microscope type OLYMPUS BX51 with software OLYMPUS Stream Motion was applied to measuring the width of the kerf on the plate. The obtained results were averaged based on 6 measurements, and are presented in Table 4.

Table 4. The matrix of experimentation based on the Box–Behnken design, with the settings of the input parameters and determined values of process performance measures.

Std	Run	T_{on} [µs]	T_{off} [µs]	DFP [MPa]	MRR [mm³/min] (Response1) *	KW [mm] (Response 2) *	Ra [µm] (Response 3) *
1	1	0.6	6	0.5	16.817	0.33	2.48
10	2	0.9	14	0.3	16.842	0.35	2.84
15	3	0.9	10	0.5	21.952	0.35	3.16
9	4	0.9	6	0.3	17.449	0.34	2.78
8	5	1.2	10	0.7	30.341	0.36	3.54
11	6	0.9	6	0.7	27.510	0.35	2.76
4	7	1.2	14	0.5	24.768	0.36	3.58
2	8	1.2	6	0.5	22.200	0.37	3.10
12	9	0.9	14	0.7	24.426	0.34	2.94
7	10	0.6	10	0.7	14.042	0.35	2.22
6	11	1.2	10	0.3	20.064	0.38	3.58
5	12	0.6	10	0.3	14.140	0.35	2.64
14	13	0.9	10	0.5	23.242	0.36	2.97
13	14	0.9	10	0.5	23.386	0.36	2.88
3	15	0.6	14	0.5	10.612	0.35	2.28

* Averaged values of measurements (for KW and Ra), or averaged calculated values for MRR.

After the samples were cut off, roughness measurements were made using the Hommel Tester T100 device. Measurements were taken along and across the cutting direction. Roughness measurements were made under the following conditions: Filter ISO11562 (M1), Pick-up TKU300, Lc (cut-off) = 0.8 mm, Lt = 4 mm, Vt = 0.5mm/s, Lc/Ls = 100. Several parameters were measured, while Table 4 contains the obtained Ra parameter values.

The present investigation focused on studying the effects of the selected variable machining parameters on the three performance measures (responses) of the WEDM process of Inconel 617 material, such as: volumetric material removal rate (MRR), dimensional accuracy (reflected by the kerf width—KW) and surface roughness (Ra).

The response surface methodology (RSM) was employed in the designing of this research experiment (DoE), for the purpose of establishing the relationship between various input process parameters and exploring the effect of these process parameters on selected output responses [1,14]. As boundaries for the experimental work were determined in-prior in the frame of a preliminary tests, the RSM method was applied in here, along with the Box–Behnken design. At the same time, the design scheme is still considered to be more proficient and most powerful than other designs, such as: the three-level full factorial design or central composite design (CCD) [20]. This concerns, in particular, the possibility for the avoidance of combined factors extremes that might deteriorate the quality of data derived from a specific experiment [14,21]. Moreover, the principal advantage of this scheme applied to the specific research case was a significant reduction in the size of experimentation (the reduced number of required process runs) to successfully examine the effect of determined input variables upon the assumed performance responses, and validate the experimental results without loss of accuracy [11,14,21].

Consequently, the DoE matrix, based on the selected design scheme, with the run order of experiments (performed on a random basis) and corresponding results of three response variables, including MRR, KW and SR (by Ra), is shown in Table 4. As it can be noted, the entire experimentation (research task) involved totally 15 runs (trials), considering three various input factors defined at three levels, instead of $3^3 = 27$ trials (in each cycle of a multiple test), as required when choosing the design of a full-scale experiment.

3. Results and Discussion

Statistical and mathematical techniques available under the RSM approach have been used to determine the correlations between the measured output responses and accepted independent process parameters. Furthermore, the ANOVA tests were performed in order to evaluate the adequacy and the significance of the developed regression model, as well as the significance of individual model coefficients (selected process parameters). Full quadratic model formulations were each time selected for fitting the models generated for the individual responses in terms of the 3 independent process parameters, given in their coded form. All those design and analysis activities were performed in the environment of Minitab statistical software package at the 95% confidence level [5,14]. The quality of fitting the obtained models to the measured values of the respective responses was described by determination coefficient of R-Squared and the standard deviation of the residual component—S [22].

3.1. Evaluation of Material Removal Rate

In the first instance, the experimental results for *MRR* were analyzed which in EDM expresses the productivity of realized machining process. Volumetric *MRR* was calculated by an Equation (9) as the product of achieved linear cutting rate in (mm/min) (obtained in the experiment for determined machine settings), the actual width of the kerf KW formed in the process in (mm), and the work piece thickness W_w (mm).

$$MRR = V_c{\cdot}KW{\cdot}W_w \tag{9}$$

The adequately calculated values of *MRR* are presented in Table 4.

Regression coefficients of the developed quadratic response surface model for MRR are in particular presented in Table 5, together with corresponding model summary statistics derived from ANOVA test. The coefficient of determination is relatively high and close to 100%, namely R-squared equals to 97.61%, that is desirable. The standard deviation of the residual component S is of small value, and amounts to 1.35023. According to data presented in the table, the effect of process parameters T-on, DFP, T-on × T-off, T-on × DFP, and T-on^2 on MRR is significant, since in all cases, the p-value is less than 0.05, and therein, the pulse-on time seems to have the greatest effect on the measured response.

Table 5. Estimated regression coefficients and summary statistics of adopted response surface full quadratic model for MRR.

Term	Coefficient	SE Coefficient	T	p
Constant	22.8600	0.7796	29.324	0.000
T-on [µs]	5.2202	0.4774	10.935	0.000
T-off [µs]	−0.9160	0.4774	−1.919	0.113
DFP [MPa]	3.4780	0.4774	7.286	0.001
T-on [µs] × T-on [µs]	−3.0854	0.7027	−4.391	0.007
T-off [µs] × T-off [µs]	−1.1754	0.7027	−1.673	0.155
DFP [MPa] × DFP [MPa]	−0.1279	0.7027	−0.182	0.863
T-on [µs] × T-off [µs]	2.1933	0.6751	3.249	0.023
T-on [µs] × DFP [MPa]	2.5938	0.6751	3.842	0.012
T-off [µs] × DFP [MPa]	−0.6192	0.6751	−0.917	0.401

S = 1.35023; PRESS = 128.702; R-Sq = 97.81%; R-Sq (pred) = 69.12%; R-Sq (adj) = 93.88%.

The results of ANOVA for the derived model with a breakdown into the individual terms, viz., linear, quadratic and the interaction are further presented in Table 6. It can be seen from the Table that the regression model for MRR proved to be significant at 95% confidence level, as it has a p-value of 0.001. It should also be noted in the same table that calculated value of the ratio of F = 24.85 is much more than its tabulated value $F_{(0.05;9;14)} = 2.65$. This implies the adequacy of the MRR related model at the assumed level of confidence. The calculated F-value for "Lack of Fit" of 4,21 implies that it is insignificant relative to the pure error (p-value = 0.198, so there is only 19.8% chance that a "Lack of Fit" F-value this large could occur due to noise during the WEDM process under study.

Table 6. ANOVA test results of the developed response surface model for MRR.

Source	DF	Seq. SS	Adj. SS	Adj. MS	F	*p*
Regression	9	407.711	407.711	45.301	24.85	0.001
Linear	3	321.492	321.492	107.164	58.78	0.000
Square	3	38.534	38.534	12.845	7.05	0.030
Interaction	3	47.685	47.685	15.895	8.72	0.020
Residual Error	5	9.116	9.116	1.823		
Lack of Fit	3	7.868	7.868	2.623	4.21	0.198
Pure Error	2	1.247	1.247	0.624		
Total	14	416.827				

Moreover, the residuals were examined using the normal probability diagnostic plot of residuals. As seen in the adequate plot in Figure 2, the residuals follow a straight line and it follows a normal distribution. This, in turn, implies that the assumptions for ANOVA are met and the regression analysis performed is valid.

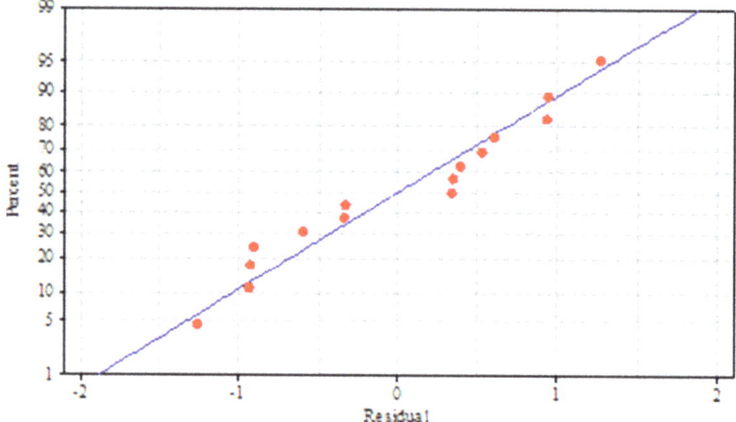

Figure 2. Normal probability plot of the residuals for MRR.

Moreover, the residuals were examined using a normal probability diagnostic plot of residuals. As seen in the adequate plot in Figure 2, the residuals follow a straight line and it follows a normal distribution. This in turn implies that the assumption for ANOVA are met and the regression analysis performed is valid.

Figure 3 shows the response surface plots related to MRR. Figure 3a, in particular, depicts the variation of MRR with pulse-on time and pulse-off time. It is seen from the graph that the productivity of the EDM process considerably increases with a combination of higher values of both input parameters. It does not apply to lower settings for T-on values, however, when increasing, T-off presents an adverse effect on MRR. Similarly, the increase in the MRR can be observed with a combination of higher T-on and higher flushing pressure DFP, as shown in Figure 3b. Figure 3c shows, in turn, that generally higher MRR can be obtained for a combination of higher DFP settings and lower T-off settings. As it can be also noted, increasing the value of T-off in process settings generally leads to a decrease in the productivity of the WEDM process of the Inconel 617 alloy.

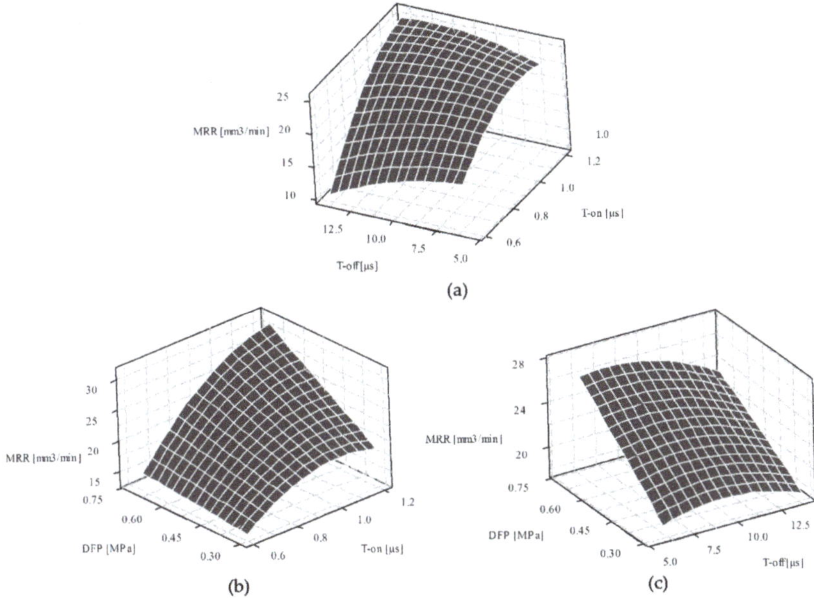

Figure 3. Response surface 3D plots representing the MRR dependence on: (**a**) T-on and T-off, for DFP = 0.5 MPa; (**b**) T-on and DFP, for T-off = 10 μs; (**c**) T-off and DFP, for T-on = 0.9 μs.

3.2. Evaluation of the Dimensional Accuracy of the Machining Process

The width of the kerf KW formed in the EDM-machined work material by a passing wire can be a measure of the dimensional accuracy of the process [11,23]. Figure 4 shows exemplary samples of kerfs made in Inconel 617 material. The averaged values of measurements for kerfs obtained in different machining conditions (correspondingly to individual runs of the experiment) are included in the matrix of Table 4.

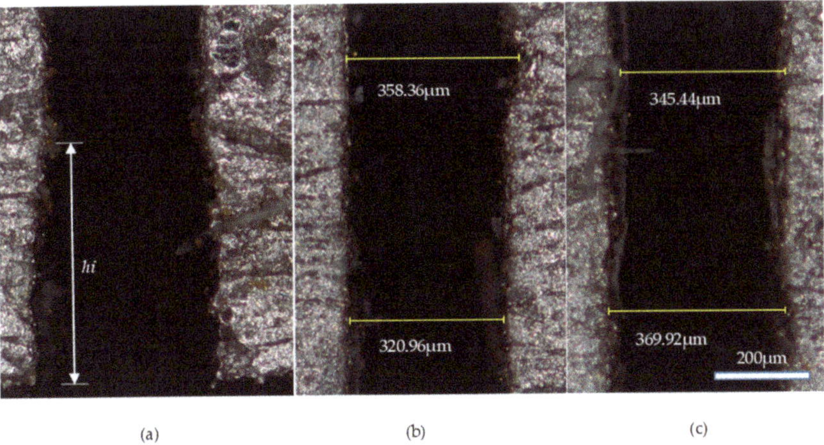

Figure 4. The photographs used in the measurements of kerf width KW: (**a**)—kerf with initial zone, (**b**)—kerf for run 1, (**c**)—kerf for run 5, h_i—depth of initial zone.

The width of the kerf can be defined by the equation:

$$KW = d_w + 2G_s \tag{10}$$

where: KW—kerf width (mm), d_w—diameter of wire (mm), G_s—lateral spark gap (mm).

As the value of wire diameter (d_w) is constant, the change of the kerf width (KW) results in changes of width lateral spark gaps (G_s). The distance of lateral sparks gap is dependent on the energy of the discharges and affects the process accuracy. Figure 4 shows the photos of the kerfs made during the tests. It can be seen that the edges of the kerfs are fuzzed and measurement of the kerf width might be inaccurate. In the initial section of kerfs (Figure 4a), the machining process is rather unstable. Thus, the section was excluded from the measurements. Figure 4b shows the kerf of the run 1 with measurement lines, while Figure 4c shows the kerf for the run 5, which corresponds to the process conditions of the highest efficiency.

The adequate regression model was determined to quantify the effect of selected input parameters on this measured response of the WEDM process studied. Calculated values of regression coefficients for the individual terms of the determined response surface model for KW are listed in Table 7. According to the data contained in this Table (see the respective p-values), in addition to the parameter T-on of the linear terms, only the T-off^2 of the quadratic terms and T-on*T-off of the interaction terms have a significant effect on KW. As also seen in the table, the values of the coefficient R-squared and the standard deviation of the residual component—S, obtained in respective ANOVA test, are at the proper level.

Table 7. Estimated coefficients of the regression model and summary statistics of adopted response surface full quadratic model for KW.

Term	Coefficient	SE Coefficient	T	p
Constant	0.356667	0.003073	116.058	0.000
T-on [µs]	0.011250	0.001882	5.978	0.002
T-off [µs]	0.001250	0.001882	0.664	0.536
DFP [MPa]	−0.0025	0.001882	−1.328	0.241
T-on [µs] × T-on [µs]	0.005417	0.002770	1.955	0.108
T-off [µs] × T-off [µs]	−0.009583	0.002770	−3.460	0.018
DFP [MPa] × DFP [MPa]	−0.002083	0.002770	−0.752	0.486
T-on [µs] × T-off [µs]	-0.0075	0.002661	−2.818	0.037
T-on [µs] × DFP [MPa]	−0.005	0.002661	−1.879	0.119
T-off [µs] × DFP [MPa]	−0.005	0.002661	−1.879	0.119

S = 0.00532291; PRESS = 0.00135; R-Sq = 93.36%; R-Sq (pred) = 36.72%; R-Sq (adj) = 81.41%.

The results of respective ANOVA test presented in Table 8 show the significance of the regression model for KW (obtained p-value < 0.05), as well as its adequacy, since the respective requirement of the F-test is met (calculated F-ratio equal to 24.85, significantly exceeds its tabulated value), all at 95% confidence level. It is also to be noted that p-value for "Lack of Fit" of 0.615 implies the non-significance of the source element, as desirable. Normal probability plot of the residuals for the KW process response is presented in Figure 5. The regularities in distribution of the residuals observed in the graph confirm that the performed analysis could be validated.

Table 8. ANOVA test results of response surface model for KW.

Source	DF	Seq. SS	Adj. SS	Adj. MS	F	*p*
Regression	9	0.001992	0.001992	0.000221	7.81	0.018
Linear	3	0.001075	0.001075	0.000358	12.65	0.009
Square	3	0.000492	0.000492	0.000164	5.78	0.044
Interaction	3	0.000425	0.000425	0.000142	5.00	0.058
Residual Error	5	0.000142	0.000142	0.000028		
Lack of Fit	3	0.000075	0.000075	0.000025	0.75	0.615
Pure Error	2	0.000067	0.000067	0.000033		
Total	14	0.002133				

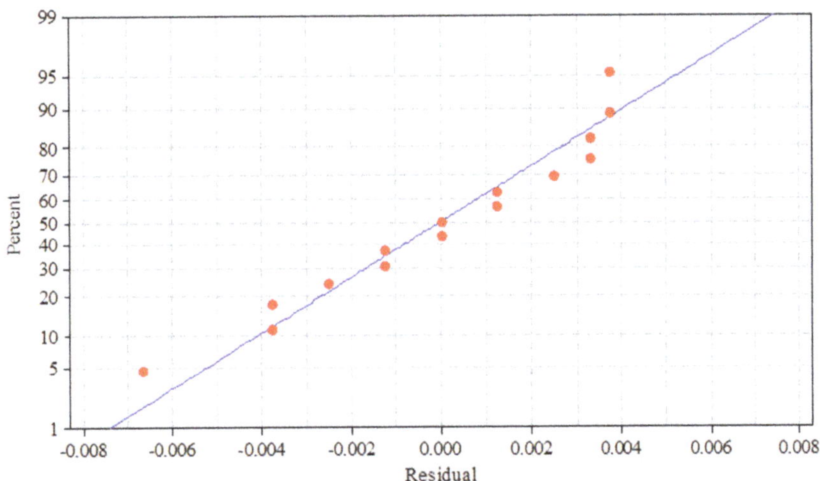

Figure 5. Normal probability plot of the residuals for KW.

Figure 6 shows the response surface 3D plots representing the KW dependence on the interacting input parameters. As generally seen, the interactions between the particular machining parameter are far more complicated than a simple rule of thumb, and their combined effect on the process response is ambiguous to much extent. Figure 6a, in particular, depicts the variation of KW with pulse-on time and pulse-off time. As can be seen in the figure, higher accuracy (i.e., smaller KW values) can basically be obtained for the combination of lower values of T-on and T-off, conversely to their effect on process productivity (the MRR response), discussed in the previous section. Figure 6b shows that the process accuracy will be higher (i.e., smaller value of KW) with a combination of smaller T-on and DFP, while a small share of the DFP parameter in the increase in the value of the measured response (KW) can be observed. The graph in Figure 6c in turn reveals that the machining accuracy increases (lower KW) for the extreme settings of the value of the T-off parameter (the lowest or the highest), as compared with its mid-settings, however, the range of observed changes in the value of this parameter is quite narrow. At the same time, a limited influence of the DFP parameter on values of the measured response can be noted.

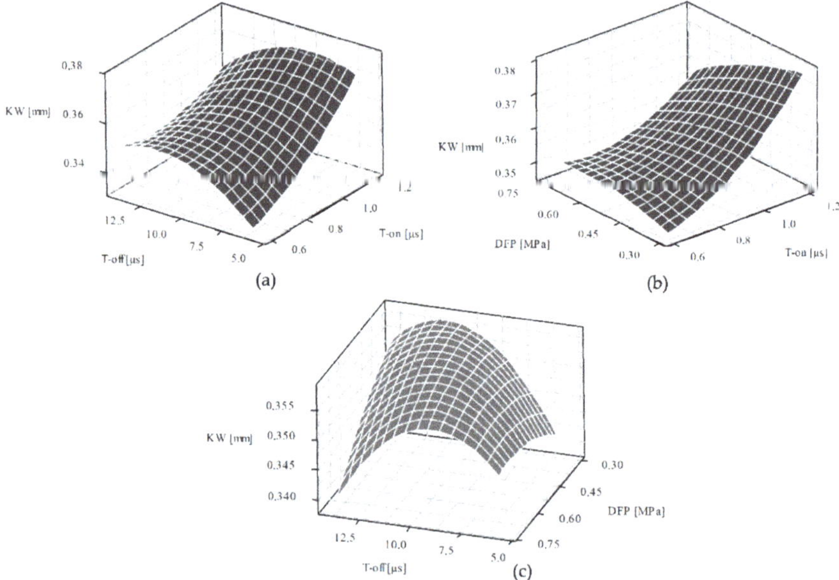

Figure 6. Response surface 3D plots representing the KW dependence on: (**a**) T-on and T-off, for DFP = 0.5 MPa; (**b**) T-on and DFP, for T-off = 10 μs; (**c**) T-off and DFP, for T-on = 0.9 μs.

3.3. Evaluation of Surface Roughness

Many of the mechanical and physical properties of mechanical parts, such as corrosion resistance, friction, fatigue strength and loading capacity, depend on the roughness of the machined surfaces [11]. Thus, the surface roughness (SR) obtained in the WEDM process, and expressed by the average roughness height (Ra) parameter, was another machining response characteristic for which a regression model was determined. The respective values of coefficients (R-squared and S) for the developed full quadratic model, that determine the quality of its fitting, are included in Table 9, along with the appropriate coefficients related to the individual model terms. As can be concluded based on the results included in this table, solely the linear parameter of T-on is significant in controlling SR (Ra). The obtained results of ANOVA test for the probability aspect P and those related to the F-test (see Table 10) indicate the significance and the adequacy of the derived response surface model for Ra, at the 95% confidence level. In addition, a relevant normal probability plot of the residuals for this output characteristic (Figure 7) has been provided that validates this part of performed analysis.

Figure 8 contains the response surface 3D plots representing the SR dependence on the interacting input parameters. Thus, Figure 8a shows the variation of Ra parameter with the change of T-on and T-off. As seen from the graph, SR considerably decreases with the decrease in pulse duration (T-on), and in fact, regardless of changes in the T-off setting. A similar effect could be observed while analyzing the variability of the Ra with the change of T-on and DFP parameters (Figure 8b). A similar effect could be observed while analyzing the variability of SR with the change of T-on and DFP parameters (Figure 8b).

Table 9. Estimated regression coefficients and summary statistics of adopted response surface full quadratic model for Ra.

Term	Coefficient	SE Coefficient	T	*p*
Constant	3.00333	0.07830	38.356	0.000
T-on [μs]	0.5225	0.04795	10.897	0.000
T-off [μs]	0.065	0.04975	1.356	0.233
DFP [MPa]	−0.0475	0.04795	−0.991	0.367
T-on [μs] × T-on [μs]	0.01083	0.07058	0.153	0.884
T-off [μs] × T-off [μs]	−0.15417	0.07058	−2.184	0.081
DFP [MPa] × DFP [MPa]	−0.01917	0.07058	−0.272	0.797
T-on [μs] × T-off [μs]	0.17	0.06781	2.507	0.054
T-on [μs] × DFP [MPa]	0.095	0.06781	1.401	0.22
T-off [μs] × DFP [MPa]	0.03	0.06781	0.442	0.677

S = 0.135622; PRESS = 0.90955; R-Sq = 96.43%; R-Sq (pred) = 64.65%; R-Sq (adj) = 89.99%.

Table 10. ANOVA test results of response surface model for Ra.

Source	DF	Seq. SS	Adj. SS	Adj. MS	F	*p*
Regression	9	2.48117	2.48117	0.27569	14 99	0.004
Linear	3	2.2359	2.2359	0.74530	40.52	0.001
Square	3	0.0997	0.08997	0.02999	1.63	0.295
Interaction	3	0.1553	0.15503	0.5177	2.81	0.147
Residual Error	5	0.09197	0.09197	0.01839		
Lack of Fit	3	0.0511	0.0511	0.01703	0.83	0.586
Pure Error	2	0.04087	0.04087	0.02043		
Total	14	2.57313				

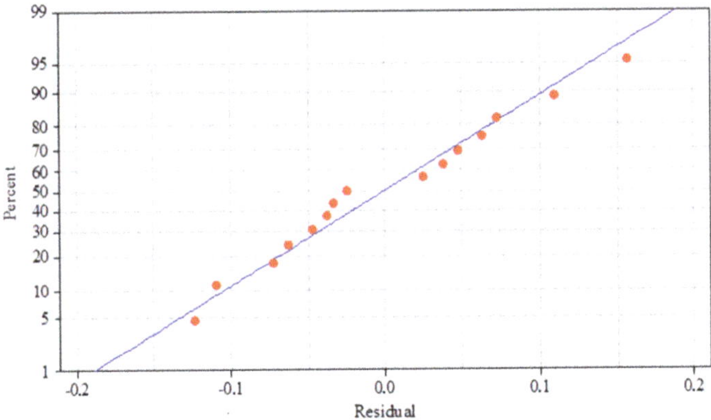

Figure 7. Normal probability plot of the residuals for surface roughness parameter of Ra.

In this instance, Ra also drops down significantly with the decrease in T-on, whereas the influence of the latter parameter is irrelevant. As can be observed in turn in Figure 8c, the dependence of the output parameter Ra on T-off and DFP is vaguer. However, for that pair of interacting parameters, the lowest values of Ra will be obtained at the lowest possible settings of the values for T-off.

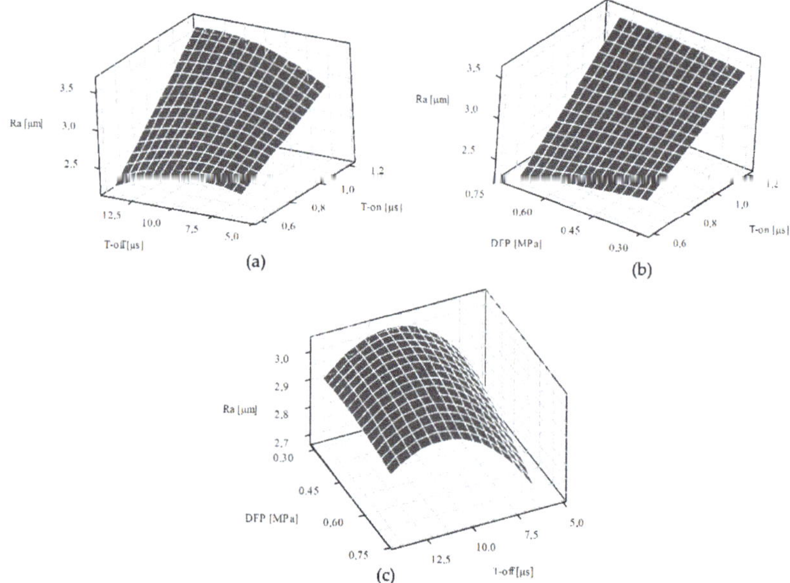

Figure 8. Response surface plots of Ra versus: (**a**) T-on and T-off, for DFP = 0.5 MPa; (**b**) T-on and DFP, for T-off = 10 μs; (**c**) T-off and DFP, for T-on = 0.9 μs.

The Ra parameter is an average value, and it is not sufficient to characterize the geometrical properties of the surface therefore the paper is supplemented by selected surface profilograms, material ratio curves and probability density functions, in order to show the operating features of obtained surfaces.

Figure 9 presents a profilogram of the surface after processing, in terms of the roughness and waviness. This presented structure applies to run 8 according to the Table 4. The roughness was filtered from the primary profile using a 0.8 mm cut off filter. Comparing the profiles shown at (a) and (b) in Figure 9, it can be seen that the roughness is the dominant element of the obtained surface structure, and from a practical point of view, the waviness in this structure is insignificant. Figure 10 shows the material ratio curves and the frequency density curve for the profiles presented in this Figure. Based on the above given curves, it can be concluded that the roughness profile unevenness structure is approximately symmetrical, i.e., the range of the peaks and valleys is comparable. The shape of material ratio curve indicates that the main cause of roughness is the EDM machining process is the characteristic of discharges. During the measurements, it was observed that the structure of unevenness slightly changes with the change of the process parameters, i.e., with certain process parameters peaks begin to dominate in the structure, and for other parameters, valley is the dominant element in the surface structure. Due to the fact that the observed changes are of very little importance for the rough machining cuts realized in this research, they have not been included.

On the other hand, by observing the waviness, and in particular, the material fraction curve and the amplitude density curve, it can be seen that the resulting waviness is influenced by several factors. These can be related to the process and the machine design features. As waviness is not the dominant factor in the surface structure, it can be neglected in the roughing cut process.

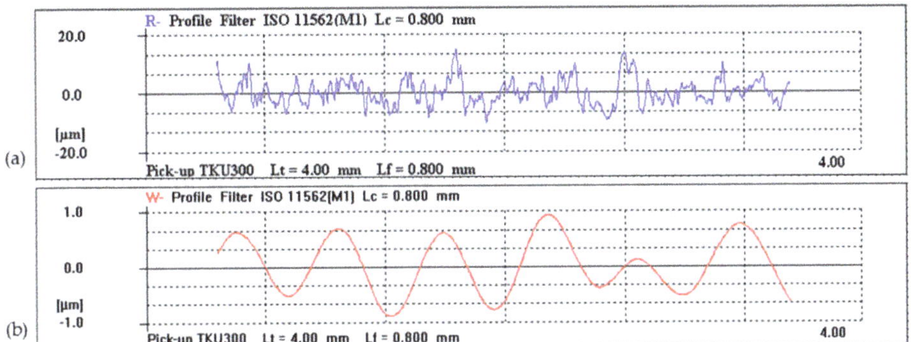

Figure 9. The surface roughness profile (**a**) and the surface waviness profile (**b**), as measured in parallel direction to the wire feed.

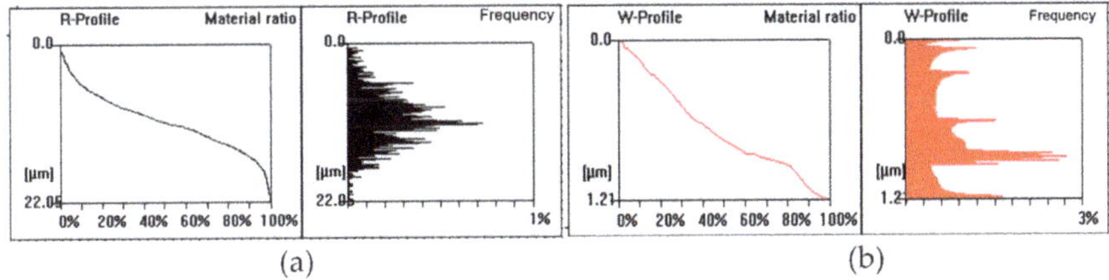

Figure 10. The material ratio curves and frequency density curves for the profiles measured in parallel direction to the wire feed: (**a**)—for the roughness, (**b**)—for the waviness.

Figures 11 and 12 show the surface structure in the direction perpendicular to the wire feed direction. Comparing these graphs with those in Figures 9 and 10, it can be seen that there are no significant differences in the roughness structure, while there are some differences in the waviness. They may result from wire vibrations during machining, whereas their influence on the surface structure effects is insignificant for rough machining.

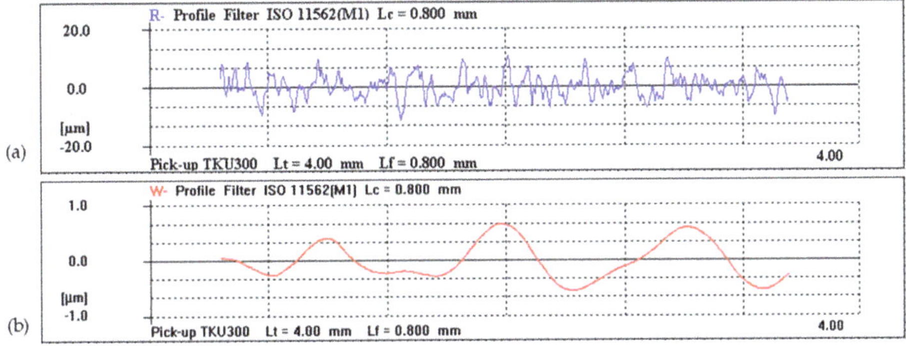

Figure 11. The surface roughness profile (**a**) and the surface waviness profile (**b**), as measured in perpendicular direction to the wire feed.

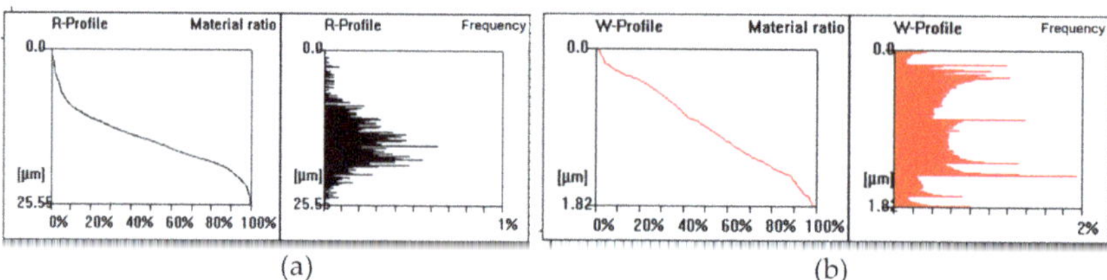

Figure 12. The material ratio curves and frequency density curves for the profiles measured in parallel direction to the wire feed: (**a**)—for the roughness, (**b**)—for the waviness.

4. Conclusions

The effect of various control parameters such as pulse-on, pulse off time, and the dielectric flow pressure on WEDM performance measures, i.e., MRR, the cutting accuracy and SR of the Inconel 617 alloy material have been experimentally investigated. The response surface modelling scheme based on the RSM approach has shown its advantages in explaining the impact of each working parameter on the values of the resultant response characteristics.

The most important results of the experimental work can be summarized as follows:

- WEDM process has generally proved its adequacy to machine Inconel 617 components in view of: the MRR values obtained in the range of 10.61–30.34 mm^3/min (65% variation in relation to the registered maximum value of MRR), the KW values—reflecting the dimensional cutting accuracy, and amounted between 0.33 and 0.38 mm (13% variation in relation to the respective maximum value) and the SR values (Ra) attained within the range of 2.22–3.58μm (38% of the recorded maximum);

- the maximum value of the MRR parameter was obtained for the following settings of the input parameters: T-on = 1.2 μs, T-off = 10 μs and DFP 0.7 MPa, maximum accuracy (KW = 0.33 mm) for T-on = 0.6 μs, T-off = 6 μs and DFP 0.5 MPa, and the smoothest surface (Ra = 2.22 μm) at T-on = 0.6 μs, T-off = 10 μs and DFP 0.7 MPa;

- T-on was found to have the greatest influence on MRR; hence, for the most efficient machining, it is necessary to ensure that the debris are removed by using effectively flushing them out (among others by increasing properly the level of DFP);

- the surface roughness structure obtained in the EDM process does not show any significant changes in the arrangement of peaks and valleys in relation to the possible settings in the variable process parameters; at the same time, a significant change concerning the waviness structure could be observed, especially for its comparisons in the longitudinal and transverse directions to the cutting direction; in the surface structure obtained in the experiment, however, waviness is not a dominant parameter and has no significant impact on its operational features.

Further research work will be focused on the investigation into the surface layer characteristics being constituted in wire EDM processes, realized under differentiated machining conditions.

Author Contributions: Conceptualization, S.D. and M.S.S.; methodology, S.D. and M.S.S.; software, S.D. and M.S.S.; validation, S.D. and M.S.S.; formal analysis, S.D. and M.S.S.; investigation, S.D. and M.S.S.; resources, S.D. and M.S.S.; data curation, S.D. and M.S.S.; writing—original draft preparation, S.D. and M.S.S.; writing—review and editing, S.D. and M.S.S.; visualization, S.D. and M.S.S.; All authors have read and agreed to the published version of the manuscript.

Funding: This research received no external funding.

References

1. Hewidy, M.; El-Taweel, T.A.; El-Safty, M. Modelling the machining parameters of wire electrical discharge machining of Inconel 601 using RSM. *J. Mater. Process. Technol.* **2005**, *169*, 328–336. [CrossRef]
2. Li, L.; Guo, Y.; Wei, X.; Li, W. Surface Integrity Characteristics in Wire-EDM of Inconel 718 at Different Discharge Energy. *Procedia CIRP* **2013**, *6*, 220–225. [CrossRef]
3. Deja, M.; Dobrzyński, M.; Flaszyński, P.; Haras, J.; Zieliński, D. Application of Rapid Prototyping Technology in the Manufacturing of Turbine Blade With Small Diameter Holes. *Pol. Marit. Res.* **2018**, *25*, 119–123. [CrossRef]
4. Sharma, P.; Chakradhar, D.; Narendranath, S. Evaluation of WEDM performance characteristics of Inconel 706 for turbine disk application. *Mater. Des.* **2015**, *88*, 558–566. [CrossRef]
5. Subrahmanyam, M.; Nancharaiah, T. Optimization of process parameters in wire-cut EDM of Inconel 625 using Taguchi's approach. *Mater. Today Proc.* **2020**, *23*, 642–646. [CrossRef]
6. Li, L.; Li, Z.Y.; Wei, X.T.; Cheng, X. Machining Characteristics of Inconel 718 by Sinking-EDM and Wire-EDM. *Mater. Manuf. Process.* **2014**, *30*, 968–973. [CrossRef]
7. Shabgard, M.; Farzaneh, S.; Gholipoor, A. Investigation of the surface integrity characteristics in wire electrical discharge machining of Inconel 617. *J. Braz. Soc. Mech. Sci. Eng.* **2016**, *39*, 857–864. [CrossRef]
8. Guo, Y.B.; Li, W.; Jawahir, I.S. Surface integrity characterization and prediction in machining of hardened and difficult-to-machine alloys: A state-of-art research review and analysis. *Mach. Sci. Technol.* **2009**, *13*, 437–470. [CrossRef]
9. Newton, T.R.; Melkote, S.N.; Watkins, T.; Trejo, R.M.; Reister, L. Investigation of the effect of process parameters on the formation and characteristics of recast layer in wire-EDM of Inconel 718. *Mater. Sci. Eng. A* **2009**, *513*, 208–215. [CrossRef]
10. Gołąbczak, M.; Maksim, P.; Jacquet, P.; Gołąbczak, A.; Woźniak, K.; Nouveau, C. Investigations of geometrical structure and morphology of samples made of hard machinable materials after wire electrical discharge machining and vibro-abrasive finishing. *Mater. Werkst.* **2019**, *50*, 611–615. [CrossRef]
11. Mostafapour, A.; Vahedi, H. Wire electrical discharge machining of AZ91 magnesium alloy; investigation of effect of process input parameters on performance characteristics. *Eng. Res. Express* **2019**, *1*, 015005. [CrossRef]
12. Klocke, F.; Schwade, M.; Klink, A.; Kopp, A. EDM Machining Capabilities of Magnesium (Mg) Alloy WE43 for Medical Applications. *Procedia Eng.* **2011**, *19*, 190–195. [CrossRef]
13. Markopoulos, A.P.; Papazoglou, E.-L.; Karmiris-Obratański, P. Experimental Study on the Influence of Machining Conditions on the Quality of Electrical Discharge Machined Surfaces of aluminum alloy Al5052. *Machines* **2020**, *8*, 12. [CrossRef]
14. Markopoulos, A.P.; Habrat, W.; Galanis, N.I.; Karkalos, N.E. Modelling and optimization of machining with the use of statistical methods and soft computing. In *Design of Experiments in Production Engineering (Management and Industrial Engineering)*; Davim, J.P., Ed.; Springer Intl. Publishing AG: Cham, Switzerland, 2016; pp. 39–76. [CrossRef]
15. Khan, A.R.; Rahman, M.; Kadirgama, K.; Maleque, M.A.; Ishak, M. Prediction of Surface Roughness of Ti-6Al-4V in Electrical Discharge Machining: A Regression Model. *J. Mech. Eng. Sci.* **2011**, *1*, 16–24. [CrossRef]
16. Rao, T.B.; Krishna, A.G. Simultaneous optimization of multiple performance characteristics in WEDM for machining ZC63/SiCp MMC. *Adv. Manuf.* **2013**, *1*, 265–275. [CrossRef]
17. Straka, L.; Dittrich, G. Intelligent Control System of Generated Electrical Pulses at Discharge Machining. In *Emerging Trends in Mechatronics*; Azizi, A., Ed.; IntechOpen: London, UK, 2020; pp. 1–26.
18. Sharma, P.; Chakradhar, D.; Narendranath, S. Effect of Wire Material on Productivity and Surface Integrity of WEDM-Processed Inconel 706 for Aircraft Application. *J. Mater. Eng. Perform.* **2016**, *25*, 3672–3681. [CrossRef]
19. Asgar, E.; Singholi, A.K.S. Parameter study and optimization of WEDM process: A Review. *IOP Conf. Ser. Mater. Sci. Eng.* **2018**, *404*, 012007. [CrossRef]
20. *NIST/SEMATECH E-Handbook of Statistical Methods*. Available online: http://www.itl.nist.gov/div898/handbook/ (accessed on 1 September 2020). [CrossRef]

21. Nguyen, T.-T.; Duong, Q.-D. Optimization of WEDM process of mould material using Kriging model to improve technological performances. *Sadhana* **2019**, *44*, 1–16. [CrossRef]
22. Mouralova, K.; Benes, L.; Bednar, J.; Zahradnicek, R.; Prokes, T.; Fiala, Z.; Fries, J. Precision Machining of Nimonic C 263 Super AlloyUsing WEDM. *Coatings* **2020**, *10*, 590. [CrossRef]
23. Saleh, M.; Anwar, S.; El-Tamimi, A.; Mohammed, M.K.; Ahmad, S. Milling Microchannels in Monel 400 Alloy by Wire EDM: An Experimental Analysis. *Micromachines* **2020**, *11*, 469. [CrossRef] [PubMed]

Article

Experimental Study on the Influence of Machining Conditions on the Quality of Electrical Discharge Machined Surfaces of aluminum alloy Al5052

Angelos P. Markopoulos [1,*], Emmanouil-Lazaros Papazoglou [1] and Panagiotis Karmiris-Obratański [1,2]

[1] Laboratory of Manufacturing Technology, School of Mechanical Engineering, National Technical University of Athens, Heroon Polytechniou 9, 15780 Athens, Greece; mlpapazoglou@gmail.com (E.-L.P.); pkarm@mail.ntua.gr (P.K.-O.)

[2] School of Mechanical Engineering and Robotics, AGH University of Science and Technology, Mickiewicza 30, 30-059 Cracow, Poland

* Correspondence: amark@mail.ntua.gr

Received: 20 January 2020; Accepted: 27 February 2020; Published: 2 March 2020

Abstract: Although electrical discharge machining (EDM) is one of the first established non-conventional machining processes, it still finds many applications in the modern industry, due to its capability of machining any electrical conductive material in complex geometries with high dimensional accuracy. The current study presents an experimental investigation of ED machining aluminum alloy Al5052. A full-scale experimental work was carried out, with the pulse current and pulse-on time being the varying machining parameters. The polishing and etching of the perpendicular plane of the machined surfaces was followed by observations and measurements in optical microscope. The material removal rate (*MRR*), the surface roughness (SR), the average white layer thickness (*AWLT*), and the heat affected zone (HAZ) micro-hardness were calculated. Through znalysis of variance (ANOVA), conclusions were drawn about the influence of machining conditions on the EDM performances. Finally, semi empirical correlations of *MRR* and *AWLT* with the machining parameters were calculated and proposed.

Keywords: electrical discharge machining; surface roughness; white layer formation; heat affected zone; ANOVA; aluminum alloy Al5052

---

1. Introduction

Electrical discharge machining (EDM) is one of the most extensively used non-conventional machining processes, with many applications in the modern industrial environment. The main advantage of EDM is the capability of machining any electrical conductive material, regardless of its mechanical properties, e.g., strength and hardness, in complex geometries and with high dimensional accuracy. EDM finds a wide range of applications in the fields of die and mold manufacturing, aerospace, automotive industries, micro-electronics, and biomedical engineering [1]. EDM is a thermoelectric process, as the material erosion mechanism is the result of a series of discrete electrical discharges. These occur between the electrode and workpiece, which are immersed in a dielectric fluid [2]. The electrical energy turns into thermal, generating a plasma channel between the cathode and the anode. With plasma temperatures in the range of 8000 to 12,000 °C, the material is heated up to its melting and/or ablation point. When the pulsating current supply is turned off, the plasma channel breaks down, allowing the dielectric fluid to flush the molten and ablated material in the form of microscopic debris.

In contrast to the plethora of studies pertaining to EDM of steel workpieces, there are only a few works about machining aluminum and aluminum alloys. Keeping in mind that aluminum and its alloys are extensively used in the aerospace and automotive industry, and its use as a matrix in composite materials, it is, beyond the scientific interest, extremely useful to be studied as a workpiece material in EDM. Khan [3] studied and analyzed the wear of the electrode during EDM. Mild steel and aluminum were used as workpiece materials and copper and brass as the tool electrode. Among the conclusions was that during machining, the mild steel electrode undergoes more wear than in the machining aluminum. This is due to the higher thermal conductivity of aluminum that results in less energy dissipation to the electrode. Gatto et al. [4] verified the machinability of three aluminum alloys, namely Al2219 and Al7050, which are used in aeronautical applications, and Al7075, a common alloy for pre-series molds. Surface finish, dimensional and geometrical accuracy of the workpieces and the wear mechanism of the electrodes were evaluated. Radhika et al. [5] investigated the influence of pulse current (10–30 A), pulse on time (120–420 μs), and dielectric flushing pressure in EDM of aluminum hybrid composites by using the Taguchi method. The material removal rate (*MRR*), the tool wear ratio (TWR), and the surface roughness (SR) were calculated or measured. Imran et al. [6] studied the machining of Al6061 with EDM under a variety of pulse current (6–12 A) and pulse-on time (60–200 μs) conditions, using graphite electrode and two different dielectric mediums. *MRR, TWR,* surface integrity in terms of the average white layer thickness (*AWLT*), and the SR were calculated. Finally, Kandpal et al. [7] and Selvarajan et al. [8], through their review papers, show the increasing interest in machining metal matrix composites (MMC), including those with an aluminum matrix, with EDM.

The current study is an experimental investigation of the effect of machining parameters, namely the pulse current I$_P$ and the pulse on time T_{on}, on the machining of aluminum alloy Al5052 with EDM. The choice of the alloy is based on its properties, as it has very good corrosion resistance to seawater and industrial atmosphere, and is used for metal matrix composites fabrication. [9] Furthermore, the working conditions under EDM processing of this material have not been thoroughly investigated, yet. In the analysis, *MRR* is calculated, Ra and Rt are measured, as indications of the machined surface quality, and through microscopic observation, *AWLT* and White Layer (WL) morphology are estimated. Finally, after the experimental procedure, non-linear regression models are calculated and proposed, correlating the *MRR* and the *AWLT* with the machining parameters, namely, the pulse current and the pulse on time.

2. Experimental Procedure

An ANGIETRON EMT 1.10 die sinking machine was used for the experiments. The workpieces were two rectangular blocks of aluminum alloy Al5052, and on each face, in successive positions, four experiments were conducted. The chemical composition of Al5052 alloy and its physical properties are presented in Table 1. A rectangular copper electrode with dimensions of 38 × 23 mm was used as the working tool, which was properly cleaned between experiments to avoid any depositions that may affect the results. A rectangular copper electrode with dimensions of 38 × 23 mm was used as the working tool, and was cleaned properly between experiments to avoid any depositions that could affect the results. Finally, a low viscosity hydrocarbon mineral oil (Castrol SE FLUID 180) was used as the dielectric fluid. Its low viscosity ensured sufficient cooling and debris flushing, even for narrow spark gaps. This dielectric fluid is recommended for machining with EDM all ferrous and non-ferrous metals.

Table 1. Chemical composition and physical properties of aluminum alloy Al5052.

Si (max)	Fe (max)	Cu (max)	Mn (max)	Mg	Cr	Zn (max)	Others (max)	Density (g/mm^3)	Hardness (HV)
0.25%	0.40%	0.10%	0.10%	2.2–2.8%	0.15–0.35%	0.1%	0.15%	0.00268	98

A full-scale experiment was carried out, with varying parameters, the pulse current I$_P$ and the pulse-on time T_{on} at four levels for each one. The 16 sets of the experiments' parameters are analytically presented in Table 2. In all experiments, straight polarity was used, with 100 and 30 V as the open and close circuit voltage, respectively. Moreover, the dielectric fluid flushing pressure was constant for all the experiments, while the duty cycle was automatically adjusted by the machine, for optimized machining efficiency. The workpiece weight before and after machining, the total machining time, the ammeter indication of the mean current intensity $\bar{I}_P$, and the machining efficiency (f_{eff}) were transcribed for subsequent calculations, evaluation, and analysis. For a more accurate calculation of the real machining time tm, the "jumping cycles", namely the periodic movement of the servo head for facilitating the debris removal during machining, were considered. Furthermore, the average and the maximum surface roughness Ra and Rt, respectively, were measured, calculated as the mean from five measurements made on each machined surface. For the values of the machined surfaces, the proper stylus tip, and the cut-off wavelengths were adopted according to EN ISO 4287:1998. Specifically, the stylus tip was 10 µm, and the shortest and longest cut-off wavelength was 2.5 and 8 mm respectively. The measuring system Mitutoyo SV-4500 CNC was used.

Finally, the machined surfaces in their perpendicular plane were grinded, polished, and chemically attacked with etchant, which was composited by 92 ml distilled water, 6 ml nitric acid and 2 ml hydrofluoric acid, and immersion of samples for 20 s in the bath. The etched surfaces were observed in optical microscope and the *AWLT* was calculated, while observations about the WL quality and morphology were made. The *AWLT* was calculated by dividing the area of the WL with its corresponding length. At the end, the micro-hardness of the material, just below the WL, was measured in order to establish the existence of a heat affected zone (HAZ). The micro-hardness was calculated as the mean from six measurements at each specimen using the Vickers hardness test.

Table 2. Experimental parameters and results.

#	I_P (A)	T_{on} (µs)	Duty Factor	MRR (mm³/min)	Ra (µm)	Rt (µm)	*AWLT* (µm)
1	15	100	0.6	173	10.8	77.2	18
2	18	100	0.61	207	10.7	76.6	17
3	21	100	0.52	207	11.2	76.2	20
4	24	100	0.54	257	11.8	86.4	21
5	15	200	0.67	180	15.1	95.6	23
6	18	200	0.66	213	14.5	98	26
7	21	200	0.58	252	14.4	96.8	27
8	24	200	0.54	274	14.5	101.6	28
9	15	300	0.65	155	14.1	93.8	36
10	18	300	0.64	217	16	113.2	37
11	21	300	0.57	224	15.2	104.2	35
12	24	300	0.58	259	14.4	104.4	37
13	15	500	0.73	177	14.2	94.4	39
14	18	500	0.72	224	17.1	114.8	42
15	21	500	0.62	234	16.7	105	42
16	24	500	0.63	280	19.4	140.6	49

The *MRR* was calculated by using Equation (1):

$$MRR = \frac{W_{st} - W_{fin}}{\rho \cdot t_m},\tag{1}$$

with *MRR* the material removal rate in mm³/min, $W_{st.}$, W_{fin} the workpiece weight before and after machining, respectively, in g, ρ the workpiece material density in g/mm³, and tm the machining time in min.

The duty cycle (η) is calculated by using Equation (2):

$$\eta = \frac{\overline{I_P}}{I_P},\tag{2}$$

with $\overline{I_P}$ the ammeter indication of the mean current intensity in A, and I_P the pulse current in A. The experimental results are also shown in Table 2.

3. Results and Discussion

3.1. Material Removal Rate

The results of Table 2 and the respective diagrams of Figure 1 indicate that the *MRR* is strongly depending on the nominal machining power P_{nom}, namely the pulse-on current I_P, as the machining voltage is constant. In general, higher pulse current leads to higher *MRR*, while the way that the pulse-on time affects *MRR* is not clear, as can be seen in Figure 1a. Thus, the real interest focuses on the interaction of machining parameters, as it is presented in the interaction plot of *MRR*, see Figure 1b. For example, the *MRR* for I_P = 15 A and T_{on} = 300 µs is lower than this for I_P = 15 A and T_{on} = 100 µs. Moreover, the *MRR* for I_P = 15 A is almost the same for T_{on} equal to 100, 200, and 500 µs. In contrast, for I_P = 21 A the maximum *MRR* has been measured for T_{on} = 200 µs, while the *MRR* for 300 µs is higher than that for 100 µs; this is the exact opposite in relation with the 15 A pulse current. From the aforementioned, it can be concluded that the interaction between the machining parameters is far more complicated than a simple rule of thumb, like higher pulse current and/or pulse-on time leads to higher *MRR*.

One of the main interests about EDM, is the energy consumption, as indication of the machining efficiency and the economic feasibility of the process. The energy consumption is analog to the mean machining power, which is calculated by using Equation (3):

$$P_{av} = P_{nom} \cdot \eta = V_P \cdot I_P \cdot \eta,\tag{3}$$

with P_{av} the mean machining power in W, and P_{nom} the nominal machining power in W.

As it is presented clearer in Figure 2a, the higher mean machining power does not always lead to higher *MRR*. Moreover, for the same mean power, the *MRR* depends on the machining parameters combination. For example, for *MRR* of approximately 175 mm 3/min, using T_{on} = 100 µs, a mean power of 270 W is needed when for the same *MRR*, using T_{on} = 500 µs, 22.2% higher mean power is needed, i.e. 330 W. At the same time, for a mean power of 390 W, the *MRR* varies from 224 to 274 mm 3/min, i.e. a 22.3% increase. As a conclusion, for the removal of the same material quantity, the required energy depends on the machining parameter combination, namely the pulse-on current and the pulse-on time, as these are the major parameters which define the power density and the per-pulse energy during the machining.

The explanation of this "behavior" is given by understanding the material removal mechanism. There are two main factors which determine the process' efficiency in EDM: the debris concentration and the gassy environment in which the spark occurs. A portion of the discharge energy is spent on re-melting and evaporating the debris material, and as a result, the efficiency is decreased. The same result, i.e., efficiency decrease, comes by the fact that a portion of the discharge takes place in a gaseous environment, a phenomenon that Gostimirovic et al. [10] pointed out in their study, too. Thus, high discharge power, which creates high debris concentration, or/and high pulse-on time, which results in a gassier environment, can decrease the efficiency of the process, with the corresponding influence on the *MRR*.

As an overall conclusion, and as a rule of thumb, it can be said that the increase in the pulse-on current results in an increase in *MRR*, while the pulse-on time has a more fuzzy and minor effect on *MRR*. As the main effect plot indicates, and as the interaction plot confirms, higher *MRR* can be achieved with higher nominal machining power, namely, higher pulse-on current.

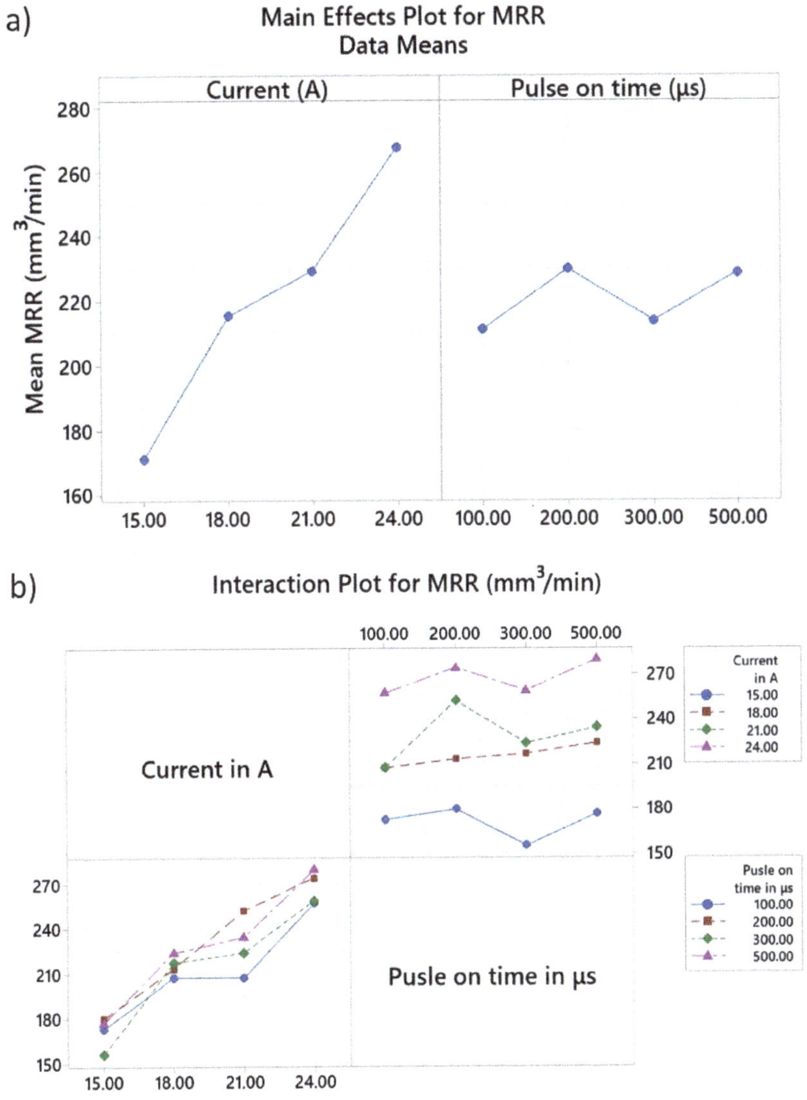

Figure 1. (**a**) Main effects plot for *MRR*. (**b**) Interaction plot for *MRR*.

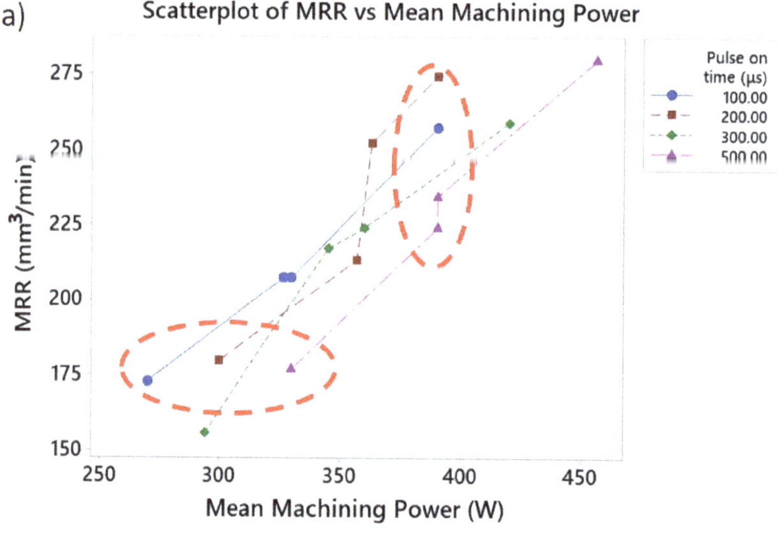

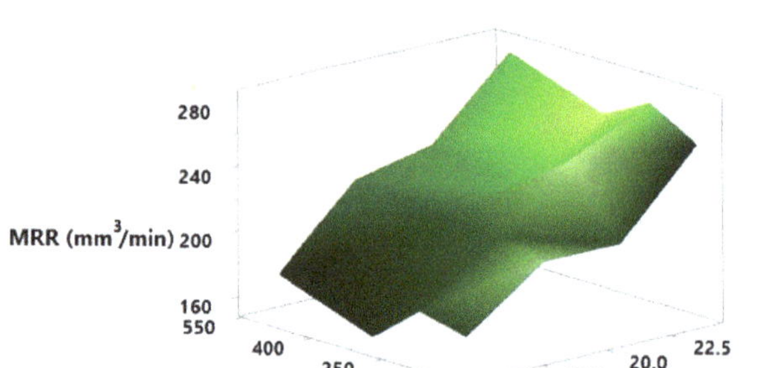

Figure 2. (**a**) *MRR* vs. mean machining power. (**b**) *MRR* vs. I_P and T_{on}.

From the experimental data and assuming a nonlinear relation, Equation (4) emerged, where *MRR* is expressed as function of I_P and T_{on}:

$$MRR = 12.77 \cdot I_P^{0.89} \cdot T_{on}^{0.038},\qquad(4)$$

with *MRR* in mm³/min, I_P in A, and T_{on} in µs. The above equation indicates that the pulse current I_P is a far more significant factor for *MRR* in contrast with the pulse-on time, as its exponent is about 20 times higher; this is confirmed in the diagram of Figure 1a.

3.2. Surface Roughness

The surface roughness depends on the geometrical characteristics of the produced craters, on the machined surface. The average craters' volume, among other parameters, as the material's thermal properties and dielectric liquid composition, depends on the discharge pulse energy. Pulse-on time and pulse current affect the crater's geometry in a different way. A simplified approach is that an increase in the pulse-on time allows the discharge channel to expand, and results in relatively big craters, while pulse current mainly reflected the depth of the produced craters [1].

The main effects plots for Ra and Rt, depicted in Figure 3a,b, respectively, indicate that the change in pulse current does not have a significant effect on Ra and Rt. This observation is confirmed by 2-samples-t test for the different pulse current values, which result in p values greater than 0.05, see Table 3. On the other hand, increasing the pulse-on time from 100 µs to 200, 300, and 500 µs results in a statistically significant change on Ra and Rt. Other changes in T_{on}, e.g., from 200 to 500 µs, do not affect in a statistically significant manner Ra and Rt.

It has to be mentioned that the surface roughness is a far more complicated phenomenon. Not only factors such as pulse-on time, pulse current, and material thermal properties should be taken into consideration, but also their interactions and overlapping.

Table 3. *p* Values of 2-samples-t test for pulse current and pulse-on time.

Ra / Rt	I_P (A)			
I_P (A)	15	18	21	24
15		0.57	0.605	0.469
18	0.35		0.917	0.84
21	0.54	0.67		0.754
24	0.24	0.62	0.39	

Ra / Rt	T_{on} (µs)			
T_{on} (µs)	100	200	300	500
100		0.000	0.002	0.014
200	0.002		0.558	0.131
300	0.006	0.252		0.192
500	0.043	0.213	0.425	

3.3. White Layer and Heat Affected Zone's Micro-Hardness

During the EDM process, the formatted craters represent only a percentage of the total melted material, as only a small amount of the material is ejected from the cavity, while the rest returns to the solid state, forming a re-solidified layer. Furthermore, as a small portion of the ablated-vaporized material has remained in close proximity to the workpiece surface, it reattaches on the surface after the end of the pulse, forming a layer known as the re-deposited layer. The re-deposited layer has a porous structure and may contain elements of the electrode material and by-products of the dielectric fluid [11]. Re-solidified and re-deposited layers are referred as the white layer.

Surface cracks are the result of extreme variations in temperature and pressure during EDM. The probability of crack formation depends on the pulse duration, current intensity, work material properties, and dielectric fluid. Although crack mechanisms are complicated, it can be said as a simplified general rule that the density of surface cracks increases with increasing the pulse energy [12]. Nevertheless, the formation and growth of cracks and micro-cracks are in major dependence with workpiece material thermos-physical properties.

In Figure 4a–d, the WL, as it is observed by microscopy imaging, is illustrated. The machining parameters are for constant pulse current I_P = 18 A and different pulse-on time, namely T_{on} = 100, 200, 300. and 500 µs. Figure 5a–d depict characteristic details, in magnification, of each workpiece of Figure 4.

It is evident that the average white layer thickness and the white layer's morphological characteristics depend on the machining conditions, and the per pulse discharge energy. For T_{on} = 100 µs, depicted in Figures 4a and 5a, the WL has an extremely thin and discontinuous profile, having only a few small globule formations. For T_{on} = 200 µs, Figures 4b and 5b, the WL is still thin and discontinuous, but larger globules appear, indicating the simultaneous existence of re-condensed (smaller globules) and incomplete evaporated (larger globules) material formations, observations which are in line with the study of Imran et al. [6]. For T_{on} = 300 µs, Figure 4c, the WL profile is thicker and almost continuous, with larger globule formation, and disintegration areas, Figure 5c. For T_{on} = 500 µs, the WL is thick, continuous, with large globules and overlapping formations, as shown in Figure 5d. Finally, for all machining parameters, porosity is observed. For shorter pulses, the porosity is limited, with finer cavities. For higher values of T_{on}, due to the larger amount of trapped air, the porosity becomes more intense, forming larger cavities, as shown in Figure 5.

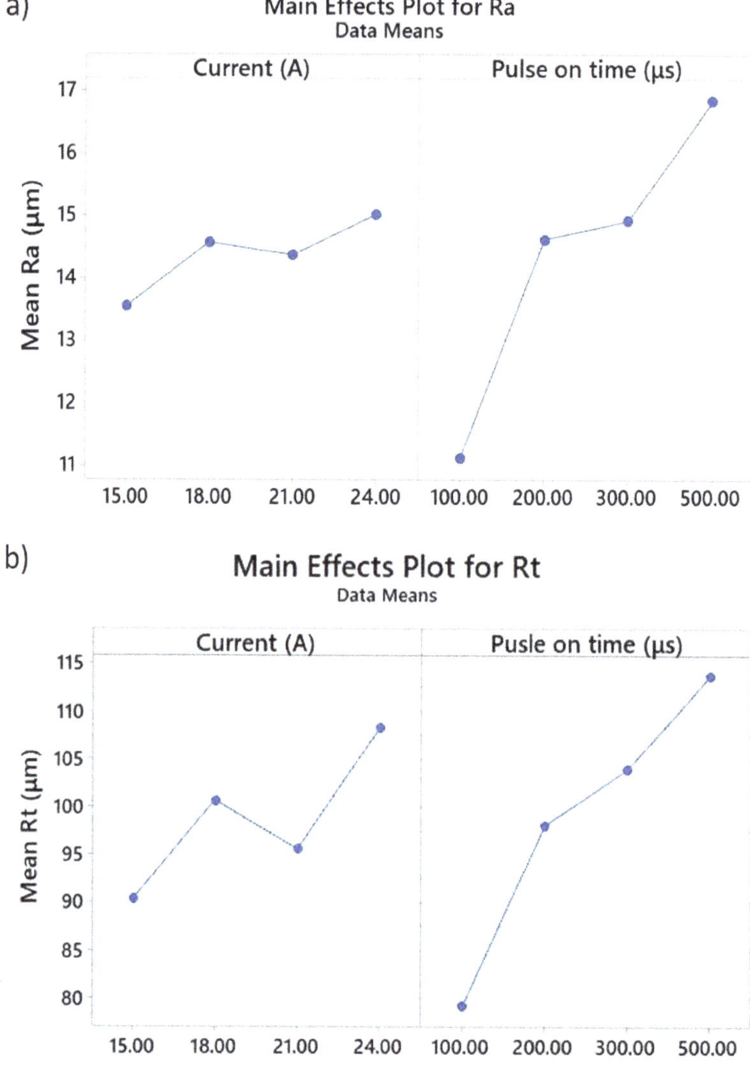

Figure 3. Main effects plot for (**a**) Ra. (**b**) Rt.

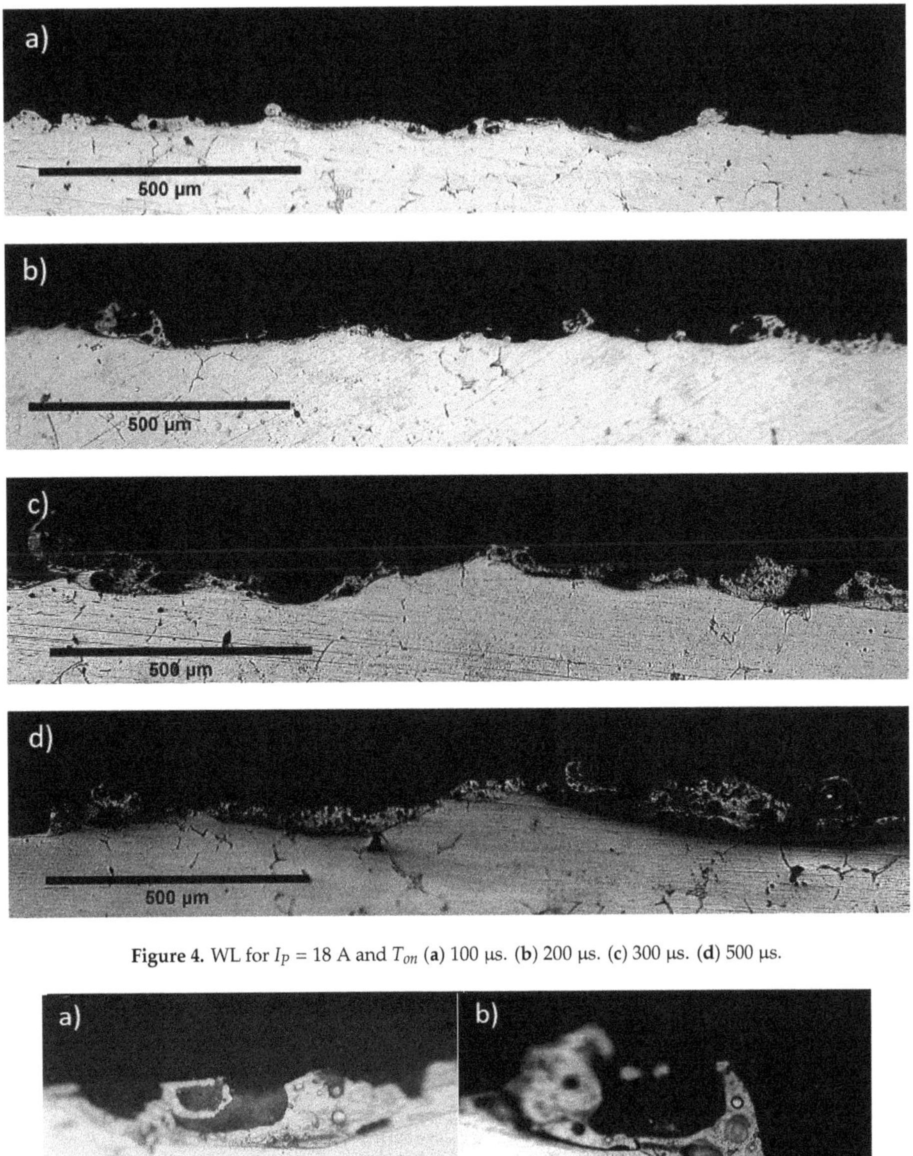

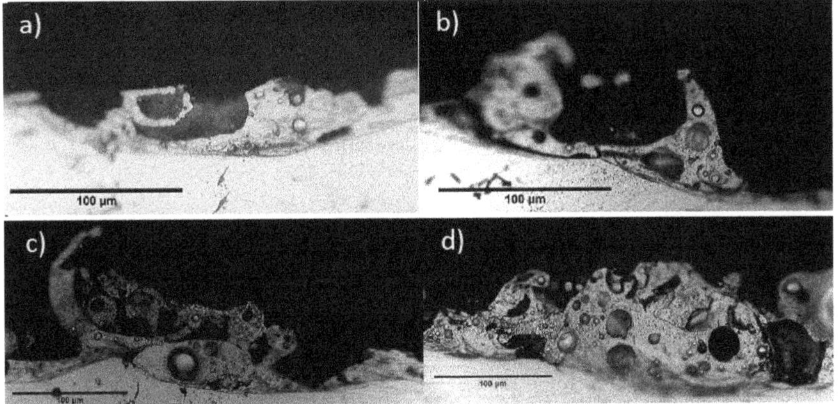

Figure 4. WL for I_P = 18 A and T_{on} (**a**) 100 µs. (**b**) 200 µs. (**c**) 300 µs. (**d**) 500 µs.

Figure 5. Details from the WL for I_P = 18 A and T_{on} (**a**) 100 µs. (**b**) 200 µs. (**c**) 300 µs. (**d**) 500 µs.

In Figure 6, the melted and re-solidified layer is depicted, along with its characteristics features, namely smaller and larger pockmarks, and islets that have been formed by molten material fluxes. It has to be pointed out the absence of intense cracks and micro-cracks on the machined surface, a typical characteristic of machined surfaces with EDM. It can be explained and attributed to the aluminum high thermal conductivity. Aluminum has a typical thermal conductivity of 138 W/mK, while steel has the half, or even lower. Thus, the thermal energy in aluminum is diffused faster, resulting in the occurrence of lower gradients of temperature, which are one of the major reasons of cracks development.

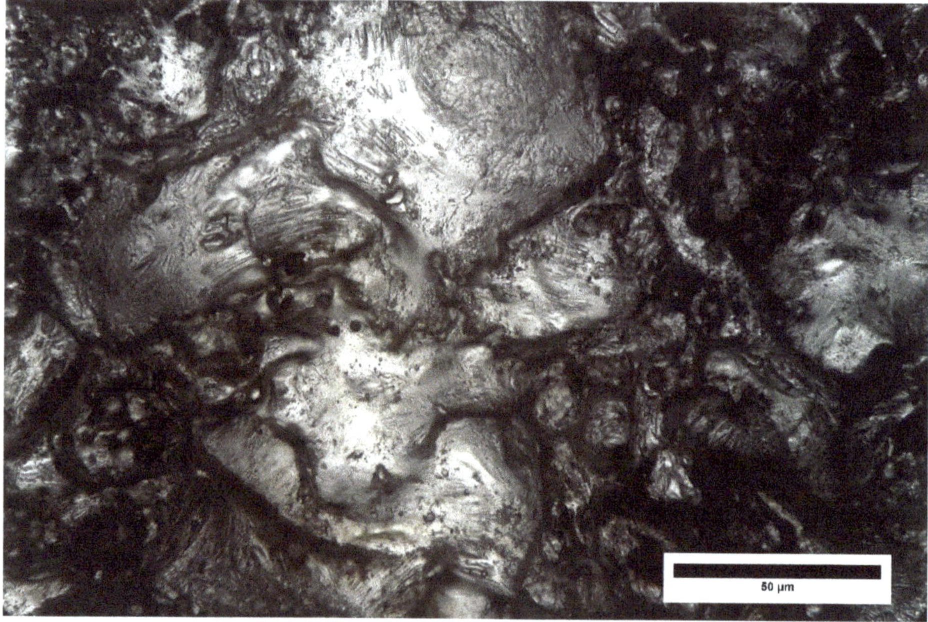

Figure 6. Details of the machined surface for I_p = 15 A and T_{on} 200 μs.

The *AWLT* and the micro-hardness of the material just beneath the WL in the HAZ were measured, and the results are presented in Table 4.

Table 4. Average white layer thickness and micro-hardness of the heat affected zone.

No. of Experiment	1	2	3	4	5	6	7	8	9	10	11	12	13	14	15	16
AWLT (μm)	18	17	20	21	23	26	27	28	36	37	35	37	39	42	42	49
Micro-hardness (HV)	63	65	50	51	50	53	50	48	52	43	48	57	57	58	58	49

As the main effect plot for the *AWLT*, shown in Figure 7a, indicates, the pulse-on time Ton has the major effect on the *AWLT* and thus, the abovementioned microscopy observations are confirmed by the measurements. Considering a nonlinear regression between the *AWLT* and the machining parameters 12 and according to the experimental results, the *AWLT* can be expressed as function of I_p and T_{on} through Equation (5):

$$AWLT = 0.743 \cdot I_P^{0.286} \cdot T_{on}^{0.52},$$

(5)

with *AWLT* in μm, I_p in A, and T_{on} in μs. Finally, the micro-hardness of the material, in the HAZ, is decreased, up to 43 HV, indicating a 37% decrease.

The formation of a HAZ and the subsequent reduction of the material's micro-hardness is the result of the thermal energy flow into the workpiece. Therefore, the higher the inflow heat energy rate is, the higher effect it has on the HAZ formation. On the other hand, the formation of a WL can have a shielding effect on the HAZ formation, by protecting the bulk material from the dissipated heat. As a result, there is no clear correlation between the machining power and the material's micro-hardness, but only evidence proving the existence of the HAZ underneath the WL, as shown in Figure 7b.

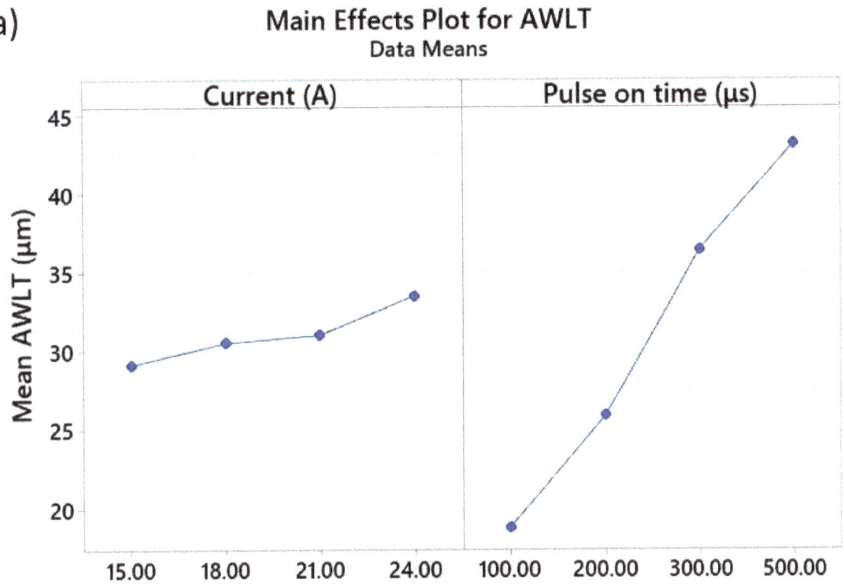

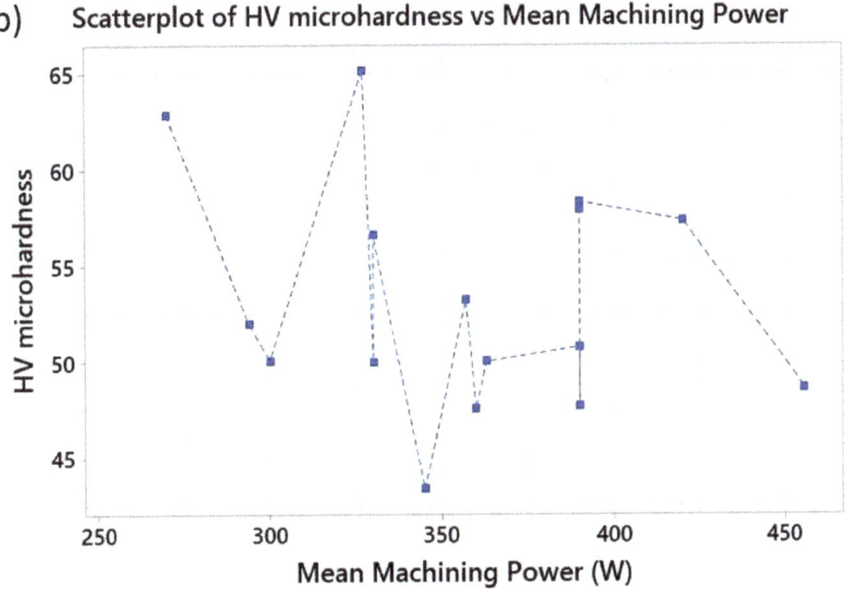

Figure 7. (**a**) Main effect plot for *AWLT*. (**b**) Scatterplot of micro-hardness HV vs. mean machining power.

4. Conclusions

An experimental investigation of machining aluminum alloy, namely Al5052, was carried out in the current study. The studied machining parameters were the pulse current I_P and the pulse-on time T_{on}. The *MRR*, the surface roughness (Ra, Rt), the *AWLT*, and the micro-hardness of the HAZ were measured and/or calculated. The most important emerged conclusions are:

- The main factor which affects the *MRR* is the pulse current I_P.
- For optimization of the machining efficiency, the interactions between machining parameters must be considered. For the same mean machining power, different *MRR* were measured; however, the same *MRR* resulted with different mean machining powers. This is the result of interactions between pulse current and pulse-on time and the manner each parameter affects the material removal mechanism.
- The surface roughness mainly depends on the pulse-on time, with the measured values having statistically significant difference, when changing from 100 µs to 200, 300, and 500 µs.
- The morphology of the WL depends on the discharge energy, mostly by the pulse-on time. Increase of the pulse energy results to a thicker WL, more continuous, with bigger globule formations and more intense porosity. The *AWLT* can be expressed as a function of I_P and T_{on}.
- The material of the HAZ has decreased micro-hardness.

Author Contributions: Conceptualization, A.P.M. and P.K.-O.; Data curation, E.-L.P.; Investigation, E.-L.P. and P.K.-O.; Methodology, A.P.M., E.-L.P. and P.K.-O.; Software, E.-L.P.; Supervision, A.P.M.; Validation, E.-L.P. and P.K.-O.; Writing – original draft, E.-L.P. and P.K.-O.; Writing – review & editing, A.P.M. All authors have read and agreed to the published version of the manuscript.

Funding: This research received no external funding.

Conflicts of Interest: The authors declare no conflict of interest.

References

1. Jahan, M.P. *Electrical Discharge Machining (EDM) Types, Technologies and Applications*; Nova Science Publishers: New York, NY, USA, 2015; p. 507.
2. Abbas, N.M.; Solomon, D.G.; Bahari, M.F. A review on current research trends in electrical discharge machining (EDM). *Int. J. Mach. Tools Manuf.* **2007**, *47*, 1214–1228. [CrossRef]
3. Khan, A.A. Electrode wear and material removal rate during EDM of aluminum and mild steel using copper and brass electrodes. *Int. J. Adv. Manuf. Technol.* **2008**, *39*, 482–487. [CrossRef]
4. Gatto, A.; Bassoli, E.; Iuliano, L. Performance Optimization in Machining of Aluminium Alloys for Moulds Production: HSM and EDM. In *Aluminium Alloys, Theory and Applications*; Kvackaj, T., Ed.; IntechOpen: Rijeka, Croatia, 2011; pp. 355–376.
5. Radhika, N.; Sudhamshu, A.R.; Chandran, G.K. Optimization of Electrical Discharge Machining Parameters of Aluminium Hybrid Composites Using Taguchi Method. *J. Eng. Sci. Technol.* **2014**, *9*, 502–512.
6. Imran, M.; Shah, M.; Mehmood, S.; Arshad, R. EDM of Aluminum Alloy 6061 Using Graphite Electrode Using Paraffin Oil and Distilled Water as Dielectric Medium. *Adv. Sci. Technol. Res. J.* **2017**, *11*, 72–79. [CrossRef]
7. Kandpal, B.C.; Kumar, J.; Singh, H. Machining of Aluminium Metal Matrix Composites with Electrical Discharge Machining - A Review. *Mater. Today Proc.* **2015**, *2*, 1665–1671. [CrossRef]
8. Selvarajan, L.; Rajavel, J.; Prabakaran, V.; Sivakumar, B.; Jeeva, G. A Review Paper on EDM Parameter of Composite material and Industrial Demand Material Machining. *Mater. Today Proc.* **2018**, *5*, 5506–5513. [CrossRef]
9. Dolatkhah, A.; Golbabaei, P.; Givi, M.K.B.; Molaiekiya, F. Investigating effects of process parameters on microstructural and mechanical properties of Al5052/SiC metal matrix composite fabricated via friction stir processing. *Mater. Des.* **2012**, *37*, 458–464. [CrossRef]
10. Gostimirovic, M.; Kovac, P.; Sekulic, M.; Skoric, B. Influence of discharge energy on machining characteristics in EDM. *J. Mech. Sci. Technol.* **2012**, *26*, 173–179. [CrossRef]

11. Jameson, E.C. *Electrical Discharge Machining*; Society of Manufacturing Engineers: Dearborn, MI, USA, 2001; p. 342.
12. Rebelo, J.C.; Dias, A.M.; Kremer, D.; Lebrun, J.L. Influence of EDM pulse energy on the surface integrity of martensitic steels. *J. Mater. Process. Technol.* **1998**, *84*, 90–96. [CrossRef]

Letter

Surface Texture after Turning for Various Workpiece Rigidities

Michal Dobrzynski and Karolina Mietka *

Department of Manufacturing and Production Engineering, Faculty of Mechanical Engineering, Gdansk University of Technology, 80-233 Gdańsk, Poland; michal.dobrzynski@pg.edu.pl
* Correspondence: karolina.mietka@pg.edu.pl; Tel.: +48-58-3471-553

Abstract: In the paper, we present an analysis of the surface texture of turned parts with L/D (length/diameter) ratios of 6 and 12 and various rigidity values. The studies were carried out on samples made of S355JR steel and AISI 304 stainless steel. A detailed analysis of 2D surface profiles was carried out by using a large number of parameters that allowed us to distinguish significant differences in the surface microgeometry, which confirmed that determining surface characteristics from one height parameter (Ra—arithmetical mean height) is far from sufficient. The obtained results indicate significantly better roughness and waviness values of the AISI 304 steel surfaces in terms of its size, periodicity, and regularity. Therefore, the turning process of AISI 304 shafts with low rigidity allows one to be able to achieve better quality texture and have a positive effect on the general properties of a workpiece. In all tested samples, surface irregularities decreased along with the distance from the tailstock. The shafts with an L/D ratio of 12 had worse surfaces in the first two sections due to lower rigidity. The results received close to the three-jaw chuck, regardless of the L/D ratio and material type, demonstrated similar waviness and roughness parameters and profiles.

Keywords: CNC turning; rigidity; surface texture; profile parameters; AISI 304; S355JR

Citation: Dobrzynski, M.; Mietka, K. Surface Texture after Turning for Various Workpiece Rigidities. *Machines* **2021**, *9*, 9. https://doi.org/10.3390/machines9010009

Received: 10 October 2020
Accepted: 7 January 2021
Published: 12 January 2021

Publisher's Note: MDPI stays neutral with regard to jurisdictional claims in published maps and institutional affiliations.

1. Introduction

During the turning process of a workpiece, cutting forces that cause an elastic deformation of a machining system—which is formed by a machine tool, grip, fixture, and cutting tool (MGFT system)—arise. The values of deformation for individual elements of this system are not constant because they depend on the applied cutting parameters and other processing conditions that exert a variable system of dynamic forces. As a result of variable cutting forces, vibrations of the machining system that significantly affect the tool path and deformation of the workpiece occur. Vibrations in the machining process are an undesirable phenomenon. They cause a number of part workmanship errors.

Since the MGFT system is dynamic and spatial, the mechanical vibrations generated in this system are also spatial. The components of vibrations in the direction perpendicular to the machined surface significantly affect the surface texture (roughness, waviness, etc.) and the components in line with the cutting speed direction-errors in the longitudinal section, e.g., conicity and a lack of parallelism between external and internal surfaces.

Deflections of element during processing comprise one of the most important factors influencing the final state of a product and, thus, its high quality expected by customers. Therefore, it is very important to optimize of the machining process to improve the dimensional accuracy of the finished product. According to Flisiak et al. [1], modelling the deflection of a workpiece during machining in terms of optimizing the processing technology of flexible elements is a priority task.

The quality of a machined surface, given by the set of roughness characteristics (microgeometry), affects the basic exploitation characteristics of the machine components [2]; therefore, surface roughness is a frequent subject of research [3–6]. For example Xavior et al. [5] investigated the effect of cutting fluids on the surface roughness of an AISI 304 steel workpiece after turning, Kaladhar et al. [6] explored the optimization of surface roughness and tool flank wear in the turning of AISI 304 austenitic stainless steel with a Chemical vapour deposition (CVD) coated tool, and Wagh et al. [7] conducted AISI 304 machinability studies using a Physical vapour deposition (PVD) system with cathodic arc evaporation (CAE) deposited onto AlCrN/TiAlN coated carbides (though they mainly analyzed the arithmetical mean height (Ra) parameter). The Ra parameter can get similar values for different tool geometries and machining principles. Different tool geometries and machining principles can lead to a surface with similar Ra parameter values. Only a detailed analysis of a 2D surface profile, described by a larger number of parameters, allows one to distinguish significant differences in surface microgeometry [2]. The authors of [8] presented the state of knowledge on the influence of surface roughness on the basic functional properties of a workpiece surface: substrate adhesion, fatigue strength, and tribological and corrosion properties. Relationships exist between 2D/3D surface roughness parameters and measurable indicators of surface functional characteristics. The authors argued that describing surface characteristics based on one height parameter (Ra), or even on several such parameters (Ra, root mean square height (Rq), maximum height of the profile (Rz), and total height of the profile (Rt)), is far from sufficient. Relationships between surface texture parameters and functional surface features should be investigated so that, on the basis of the first measurements, it is possible to predict the individual functional properties of manufactured parts.

The formation process of the geometric structure of the surface as a result of machining is complex and is influenced by many factors [9]. This process consists of the effect of individual elements of a machining system. In order to obtain specific surface texture parameters in the machining process, it is necessary to consider the kinematic and dynamic features characterizing the machine tool; the stereometry, dimensions, and properties of the tool material; and the physical properties and dimensions of the processed material.

The material qualities that influence the of surface texture are machinability (which is determined by the mechanical properties like hardness and strength), chemical composition, and material structure.

Another important feature is the rigidity of the material, which is defined as the force needed to deform an object. Material rigidity is often characterized by the Young's modulus, a value that depends on the chemical composition, crystal structure, and phase composition of the microstructure. During longitudinal turning, rigidity is also influenced by the ratio of the workpiece length (L) to its diameter (D). In order to eliminate the influence of low workpiece rigidity, appropriate clamping is used. Liu et al. [10] implemented a finite difference (FD) analysis method for calculating the deformations of multi-diameter workpieces during cutting. The authors of [11] presented a mathematical model of the system of longitudinal turning and a mathematical model of the dynamic system of the machining of shafts with a low rigidity in the elastic-deformable state while considering factors interfering with and destabilizing the process of shaft machining. Benardos et al. [12] verified that the numerical method and development of an ANN (artificial neural network) model were based on data gathered from turning experiments conducted on a CNC lathe, thus allowing for a reduction of workpiece elastic deflections under cutting forces in turning.

The authors of [13] described a machining station, working together with a lathe, designed and constructed for the stabilization of the axis of low-rigidity parts in the process of machining. The basic element was a self-centering lunette with a hydraulic drive, allowing for part centering without any preliminary alignment. Świć et al. [14] presented an analysis of the possibility of increasing the accuracy and stability of the machining of low-rigidity shafts. A way of improving the accuracy of machining of shafts was to increase rigidity as a result of the oriented change of the elastic-deformable state through the application of a tensile force which, combined with the machining force, forms longitudinal–lateral strains.

2. Materials and Methods

The materials under investigation were in the form of Ø30 mm drawn bars, with length of 393 and 213 mm, made of S355JR steel and AISI 304 stainless steel, respectively. Machining processes were carried out on a TUR-50 CNC turning machine. A workholding method based on clamping shaft in a three-jaw chuck with a distance of 33 mm and a hydraulic pressure of 1.8 MPa supported by a revolving center (with a pressure of 1.2 MPa) (Figure 1). Every process was made with a coolant (SITALA D201.03 made by Houghton). The machining parameters in the turning operation were established as the following: a feed (f) of 0.20 mm and a cutting speed (vc) of 75 m/min. The cutting depth (ap) values for the S355JR and AISI 304 shafts were 0.7 and 0.8 mm, respectively. In the research, TWLNR 2525 M08 turning knives with turning inserts of the same geometry—WNMG 080408-MA MC6025 for S355JR and WNMG 080408-MA MC7025 for AISI 304 (both made by Mitsubishi Materials)—were applied. During investigation, two L/D (length/diameter) ratios of 6 and 12 of the turned shaft were considered (Table 1).

(a)

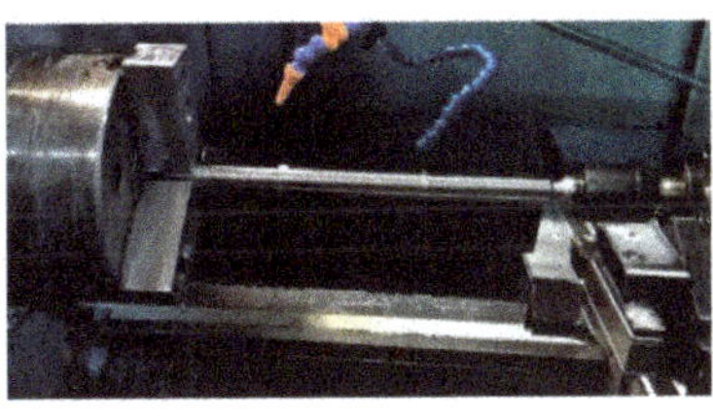
(b)

Figure 1. The view of the shafts with length/diameter (L/D) ratios of 6 (a) and 12 (b) during machining.

Table 1. Designation of samples under testing.

Material	AISI 304	S355JR
L/D = 12	S#1	S#3
	S#2	S#4
L/D = 6	S#5	S#7
	S#6	S#8

The surface texture validations were carried out with a Hommelwerk Standard 1000 surface roughness tester. During the measurements, the following parameters were adopted: evaluation length (ln) = 12.5 mm; cut off value (λc) = 2.5 mm; cut off ratio ($\lambda c/\lambda s$) = 300; and sampling interval = 1.5 µm. The stylus end was conical (taper angle of cone = 60°) with a spherical tip (tip radius (rtip) = 2 µm). These values were adopted because the main objective was to assess the surface condition on the possibly greatest length that was limited by the maximum measuring length of the used device. The shafts with an L/D of 12 were measured in four sections and the shafts with an L/D of 6 were measured in two sections at the distances presented in Figure 2. At each section, three parallel-to-the-feeding-direction measurements with a step of 120° were done. The surface texture for shafts with different rigidities was validated using the maximum height of the waviness profile (Wz) parameter, as well as selected roughness parameters. An analysis based on the Ra parameter was insufficient to define the surface texture, so the investigation considered other parameters such as Rq, Rdq (the root mean square for the local slope dz/dx within the sampling length), Rz, and Rt. A detailed analysis of the 2D surface profiles of the samples made of S355JR steel and AISI 304 stainless steel using a large number of parameters enabled us to distinguish significant differences in the surface micro-geometry and measurable indicators of surface functional characteristics.

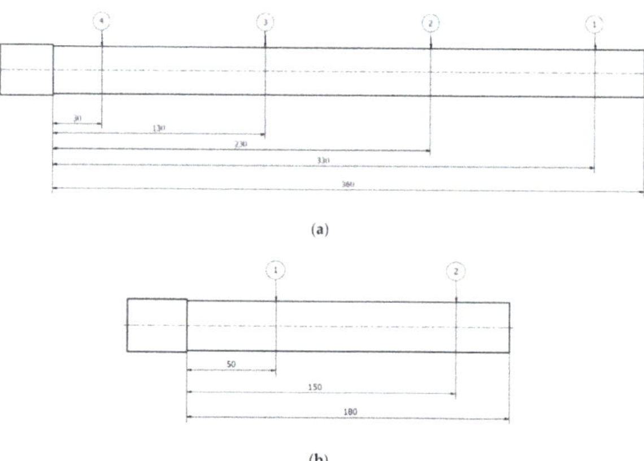

(a)

(b)

Figure 2. This the measurement positions for the shaft with L/D = 12 (**a**) and L/D = 6 (**b**).

3. Results and Discussion

Measuring the surface texture enabled a comparison of surface structures after turning AISI 304 and S355JR shafts with different L/D ratios. In comparing the profiles of the samples under investigation, significant changes in the appearance of the roughness (R) and waviness (W) profiles (the primary (P) profile was added for better visualization) could be noticed (Figures 3–6).

In the case of turning the S355JR steel, it could be seen that the surface waviness obtained higher values than when machining the AISI 304 stainless steel, and greater waviness periodicity occurred. In addition, the profiles had greater irregularity. This was due to the adoption of various workpiece rigidities (L/D ratios) and steel/stainless steel properties such as machinability. It was also noticeable that the greater the distance from the tailstock, the lower the surface waviness that appeared in all samples. This was due to the vibrations of the maximum amplitude that occurred at the beginning of machining, related to the initial wear of the blade and the distribution of machining forces. Similar phenomena were observed in the roughness profiles. The turning operations of S355JR steel shafts resulted in an increase in surface roughness, and the R profiles were more irregular (both in the horizontal and vertical directions). The surfaces farther from the tailstock showed decreasing profile micro-irregularities.

By comparing the shafts with L/D = 6 with those with L/D = 12, it could be concluded that the longer workpieces had greater waviness and surface roughness in the first two sections. This was due to the lower rigidity of the shaft in these places during machining. Meanwhile, the waviness and roughness of the surfaces of the shafts of the same material closest to the three-jaw chuck had similar shapes and resulting R and W parameter values.

When comparing the third section of the shaft with L/D = 12 with the first section of the shaft with L/D = 6, it could be seen that despite the similar distance from the three-jaw chuck, the surface waviness and roughness were greater for the shorter shafts due to the phenomena occurring at the beginning of the machining.

Based on the values of the maximum height of the Wz parameter, a significant increase in the surface waviness of the S355JR steel samples compared to the AISI 304 stainless steel samples was noted, and the maximum difference in the concerned cases was nearly 107% (Figure 7). The values of the Wz parameter were similar in the shafts on the chuck side, while they were significantly different at the end of the samples (when supported by the center) (Table 2). Shafts with L/D = 12 in the turning operation had more than twice the maximum height of the W profile in comparison with the shafts with L/D = 6 made of AISI 304 stainless steel, and these values were three times higher for workpieces made of S355JR steel. The increased height of the waviness profile from the center support caused by vibrations of maximum amplitude at this point. This was also confirmed by the lower rigidity of the support from the center side.

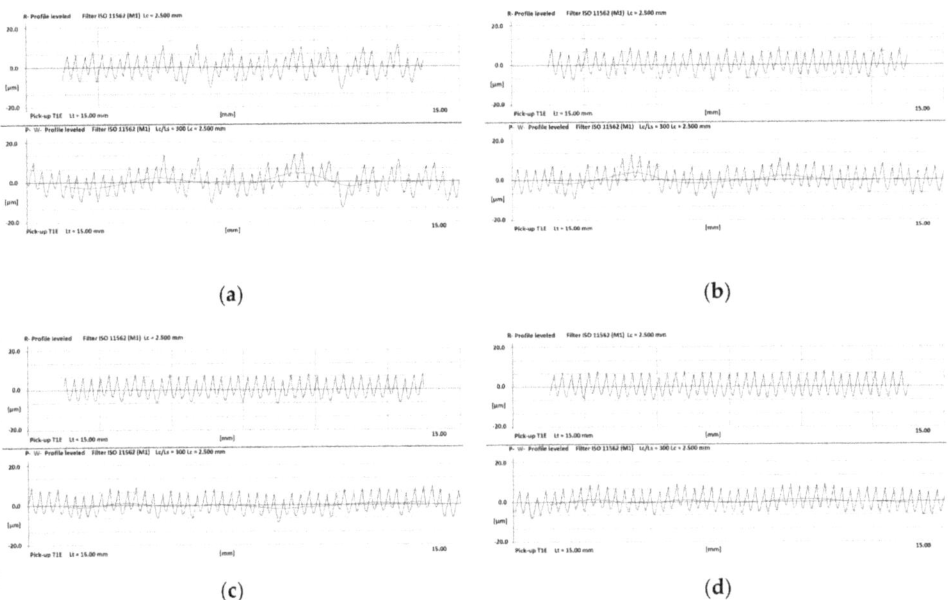

Figure 3. R (roughness), W (waviness), and P (primary) profiles of the first machined sample (L/D = 12; AISI 304): (**a**) section 1, (**b**) section 2, (**c**) section 3, (**d**) section 4.

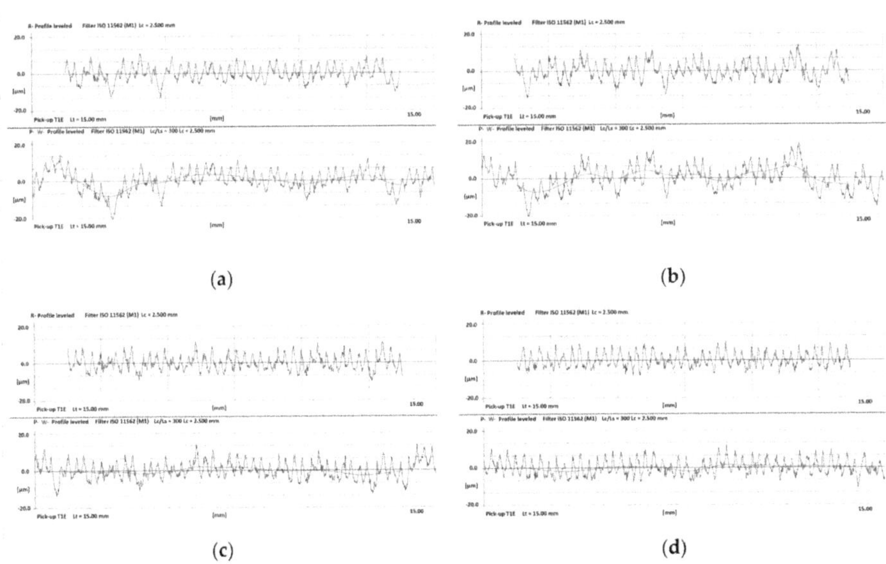

Figure 4. R, W, and P profiles of the third machined sample (L/D = 12; S355JR): (**a**) section 1, (**b**) section 2, (**c**) section 3, (**d**) section 4.

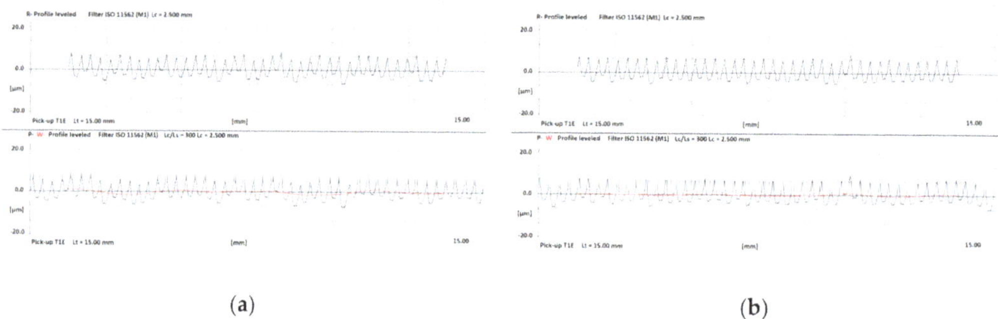

(a) **(b)**

Figure 5. R, W, and P profiles of the sixth machined sample (L/D = 6; AISI 304): (**a**) section 1, (**b**) section 2.

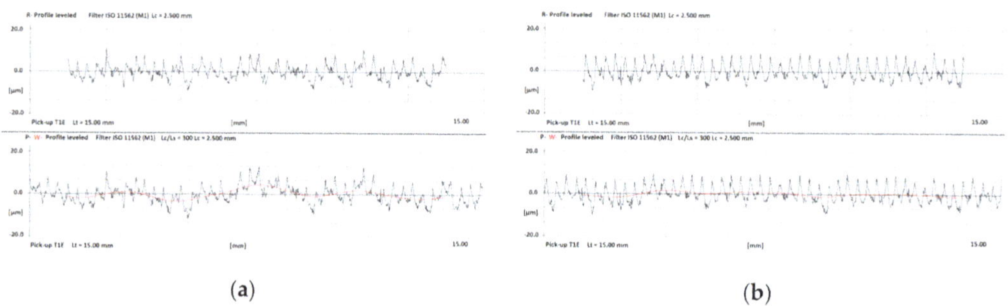

(a) **(b)**

Figure 6. R, W, and P profiles of the eighth machined sample (L/D = 6; S355JR): (**a**) section 1, (**b**) section 2.

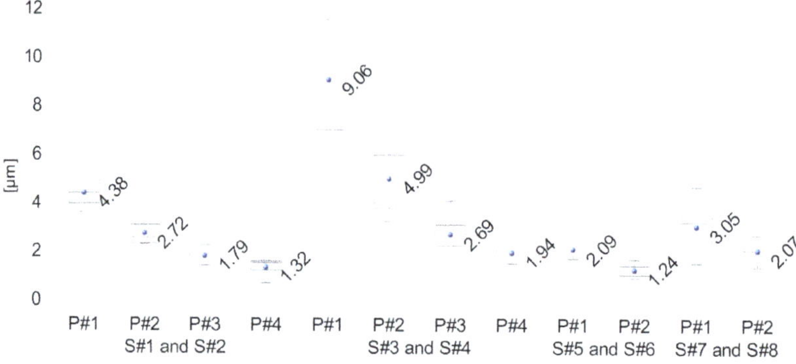

Figure 7. The center and spread of a maximum height of the waviness profile (Wz) of the samples under investigation (the box plot presents the mean, the median, the interquartile range box, and the range of the data).

Despite the higher cutting resistance, the surface of the stainless steel workpieces after turning presented better quality, and significantly lower values of R parameters were received.

The irregularities of the parts after turning were evaluated by applying mean parameters, such as Rq (Table 3) and Ra (Table 4). The values of the Ra and Rq parameters in the case of the AISI 304 stainless steel samples were characterized by a relatively small dispersion of the obtained results. On the other hand, the machining of shafts with L/D = 12 made of S355JR steel resulted in an increase in the average roughness parameters, as well as their dispersion (Figure 8). The observed individual maximum and minimum height values had a larger effect on Rq values than on Ra. In all tested cases, the Rq parameter values were higher than Ra by an average of 20.37%.

Table 2. Main statistics of the maximum height of the Wz parameter.

		Section	Mean	Median	Q1	Q3	Max	Min
		P#1	4.38	4.37	3.93	4.91	5.13	3.56
	AISI	P#2	2.72	2.53	2.29	3.06	3.57	2.26
	304	P#3	1.79	1.75	1.69	1.90	2.22	1.38
L/D = 12		P#4	1.32	1.50	1.19	1.56	1.60	0.66
		P#1	9.06	10.20	6.99	10.50	11.54	5.92
	S255JR	P#2	4.99	4.76	3.78	5.96	7.41	3.21
		P#3	2.69	2.60	2.22	3.08	4.09	1.51
		P#4	1.94	2.02	1.97	2.06	2.08	1.49
	AISI	P#1	2.09	1.93	1.91	2.07	3.00	1.69
	304	P#2	1.24	1.22	1.03	1.41	1.67	0.90
L/D = 6		P#1	3.05	3.22	1.90	3.94	4.68	1.52
	S255JR	P#2	2.07	2.07	1.66	2.54	2.68	1.36

Table 3. Main statistics of the mean square height (Rq) parameter.

		Section	Mean	Median	Q1	Q3	Max	Min
		P#1	4.42	4.40	4.14	4.68	4.83	4.03
	AISI	P#2	3.74	3.74	3.72	3.77	3.78	3.67
	304	P#3	3.58	3.59	3.55	3.61	3.63	3.53
L/D = 12		P#4	3.67	3.67	3.63	3.69	3.72	3.62
		P#1	5.49	5.79	4.44	6.27	7.03	3.91
	S255JR	P#2	4.56	4.55	3.97	5.13	5.31	3.86
		P#3	3.78	3.80	3.66	3.92	3.93	3.60
		P#4	3.72	3.71	3.69	3.76	3.94	3.52
	AISI	P#1	3.09	3.07	3.04	3.10	3.31	2.93
	304	P#2	3.32	3.32	3.31	3.34	3.46	3.17
L/D = 6		P#1	3.30	3.35	3.18	3.52	3.59	2.81
	S255JR	P#2	3.22	3.19	3.01	3.38	3.62	2.93

Table 4. Main statistics of the arithmetical mean height (Ra) parameter.

		Section	Mean	Median	Q1	Q3	Max	Min
		P#1	3.65	3.65	3.45	3.85	3.95	3.38
	AISI	P#2	3.18	3.18	3.16	3.21	3.22	3.14
	304	P#3	3.09	3.10	3.06	3.11	3.12	3.04
L/D = 12		P#4	3.16	3.16	3.14	3.18	3.20	3.13
		P#1	4.38	4.68	3.56	4.95	5.59	3.13
	S255JR	P#2	3.63	3.58	3.20	4.09	4.18	3.13
		P#3	3.11	3.12	2.99	3.22	3.28	2.94
		P#4	3.10	3.11	3.05	3.16	3.27	2.91
	AISI	P#1	2.61	2.60	2.57	2.64	2.80	2.45
	304	P#2	2.85	2.86	2.84	2.87	2.93	2.73
L/D = 6		P#1	2.65	2.70	2.59	2.83	2.85	2.23
	S255JR	P#2	2.63	2.59	2.43	2.78	2.99	2.37

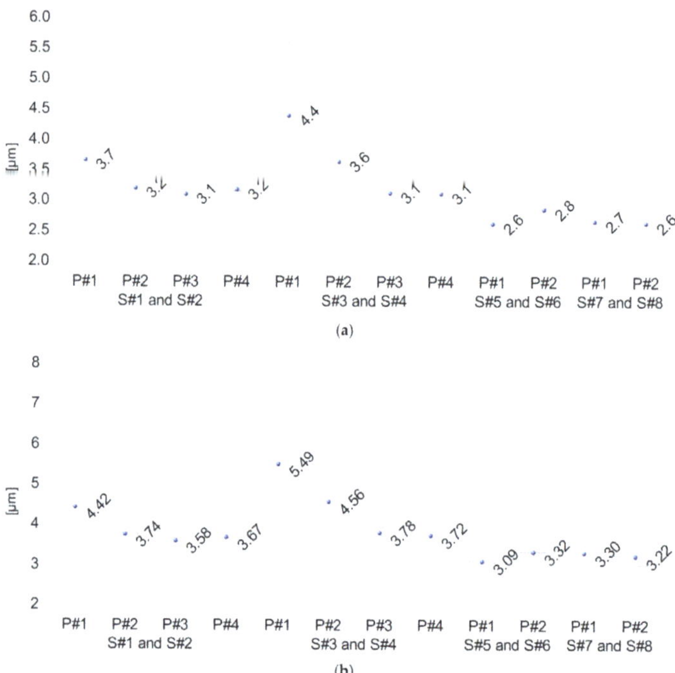

Figure 8. The center and spread of Ra (**a**) and Rq (**b**) (the box plot presents the mean, the median, the interquartile range box, and the range of the data).

The steepness of the surface could be numerically represented with the Rdq parameter, and the analysis of that parameter's values made it possible to assess the susceptibility of the tested surfaces to processing. The advantage of this parameter is its high sensitivity to extreme peaks of the profile; low values characterize smooth surfaces, and higher values characterize rough surfaces (Table 5). For all shafts made of AISI 304 stainless steel, the Rdq parameter was in the range of 0.077°–0.091° with a relatively small dispersion (Figure 9). On the other hand, shafts made of S355JR steel, regardless of the L/D ratio, were characterized by large dispersions of the Rdq parameter, with values in the range of 0.099°–0.157°. In the case of section 4 (P#4), the highest value of the Rdq parameter was obtained for the S#3 shaft, and the greatest dispersion of the parameter occurred in section 2 (P#2) for S#4 shaft (Table 5).

Table 5. Main statistics of the Rdq parameter.

		Section	Mean	Median	Q1	Q3	Max	Min
L/D = 12	AISI 304	P#1	0.089	0.090	0.088	0.091	0.092	0.085
		P#2	0.086	0.088	0.084	0.088	0.088	0.082
		P#3	0.086	0.086	0.084	0.088	0.088	0.083
		P#4	0.091	0.091	0.090	0.092	0.092	0.090
	S255JR	P#1	0.105	0.105	0.100	0.110	0.117	0.095
		P#2	0.130	0.134	0.124	0.137	0.142	0.111
		P#3	0.147	0.147	0.138	0.154	0.160	0.135
		P#4	0.149	0.150	0.141	0.152	0.168	0.136
L/D = 6	AISI 304	P#1	0.077	0.078	0.076	0.079	0.079	0.075
		P#2	0.083	0.083	0.083	0.083	0.084	0.079
	S255JR	P#1	0.117	0.118	0.111	0.124	0.129	0.105
		P#2	0.129	0.127	0.125	0.132	0.142	0.123

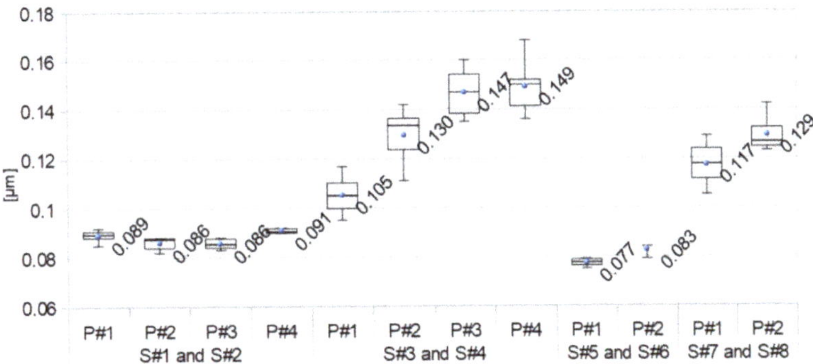

Figure 9. The center and spread of the Rdq parameter (the box plot presents the mean, the median, the interquartile range box, and the range of the data).

The values of the Rt in the considered cases were approximately 15.76% greater than the corresponding Rz values. For the S355JR steel shafts, the values of the Rz and Rt parameters were higher and had a greater dispersion. The highest Rz values, Rt values, and Rt spread were observed for the S#4/1 sample section. However, the largest spread of the Rz parameter was for the S#3/1 sample section. The main statistics of the Rz and Rt parameters of the shafts with L/D ratios of 12 and 6 are presented in Table 6.

Table 6. Main statistics of the total height of the profile (Rt) and maximum height of the profile (Rz) parameters.

			Total Height of the Profile—Rt				Maximum Height of the Profile—Rz			
		Section	Mean	Median	Max	Min	Mean	Median	Max	Min
L/D = 12	AISI 304	P#1	21.69	22.20	24.43	18.43	18.99	19.00	21.25	16.83
		P#2	16.93	16.71	18.12	16.24	15.21	15.23	15.60	14.70
		P#3	15.28	15.23	16.47	14.45	14.03	14.01	14.76	13.42
		P#4	14.77	14.76	15.21	14.31	13.92	13.89	14.20	13.66
	S255JR	P#1	30.22	29.22	39.62	23.81	23.84	25.16	27.95	17.88
		P#2	28.46	28.67	34.09	23.54	21.21	21.67	23.67	17.98
		P#3	21.73	21.81	23.94	19.02	17.82	17.70	18.85	17.08
		P#4	19.26	18.97	21.59	17.72	16.83	16.78	18.58	15.58
L/D = 6	AISI 304	P#1	13.70	13.64	15.27	12.52	12.28	12.19	13.38	11.67
		P#2	13.81	13.44	15.37	13.01	12.52	12.33	13.90	12.00
	S255JR	P#1	19.37	20.28	22.16	15.09	16.53	16.79	19.01	14.02
		P#2	17.22	16.95	19.14	16.11	15.28	15.15	17.09	14.05

All parameters had the greatest dispersions for shafts with L/D = 12 made of S355JR steel; this phenomenon was caused by the instability of the machining process, which was also visible in the R and W profiles. The occurring waviness was caused by vibration in the MGFT system, so it is likely that the cutting edge was operating under uneven conditions, which could also have affected the surface texture. Furthermore, the samples were turned from a drawn bar, which, when processed, could obtain non-uniform properties. The AISI 304 stainless steel obtained a significantly smaller spread of parameter values, regardless of the L/D ratio. The surface texture after the turning of stainless steel is periodic, which leads to positive effects on workpiece properties after machining. In all samples under investigation, surface irregularities decreased along with the distance from the tailstock. The shafts with an L/D ratio of 12 obtained worse surfaces in the first two sections, which resulted from their lower rigidity. Regardless of the L/D ratio, similar waviness and roughness profiles close to the three-jaw chuck were obtained. The Ra and Rq roughness parameters had no differences due to the material type, whereas the Rz, Rt, and Rdq parameters had higher values for the S355JR steel.

4. Conclusions

The low rigidity of workpieces in relation to the rigid parts of a machine tool hinders the cutting process due to the generated vibrations. The properties of a workpiece material crucially affect the accuracy of execution.

The above-discussed data were also related to better properties of the AISI 304 material, which is characterized by a better machinability. The roughness parameters were found to improve with distance from the tailstock due to the phenomena that occurred at the beginning of the turning, the vibrations of the maximum amplitude at the point of the support of the center, and the lower rigidity of this support compared to the clamping in the chuck. For the same reasons, the shafts with an L/D ratio of 6 obtained better surface textures. In summarizing the results, it can be stated that the rigidity of a workpiece has a large impact on the geometric structure of a surface. Generally, analyses based on the Ra parameter are insufficient to define surface texture, so this paper considered other parameters as well.

Author Contributions: Conceptualization, M.D. and K.M.; formal analysis, M.D. and K.M.; investigation, M.D. and K.M.; methodology, M.D. and K.M.; writing—original draft, M.D. and K.M.; writing—review and editing, M.D. All authors have read and agreed to the published version of the manuscript.

Funding: This research received no external funding.

Acknowledgments: The authors wish to thank the firm GloBart for providing machine tool and raw material used in the study.

Conflicts of Interest: The authors declare no conflict of interest.

References

1. Flisiak, J.; Józwik, J.; Włodarczyk, M. Analyzing the machining technology of the bolt for improvement dimensional accuracy of the finished product. *Maint. Reliab.* **2005**, *2*, 48–52.
2. Dobrocký, D.; Studený, Z.; Pokorný, Z.; Joska, Z.; Faltejsek, P. Assessment of Surface Structure of Machined Surfaces. *Manuf. Technol.* **2019**, *19*, 563–572. [CrossRef]
3. Duplák, J.; Zajac, J.; Hatala, M.; Mital', D.; Kormoš, M. Study of Surface Quality after Turning of Steel AISI 304. *Manuf. Technol.* **2014**, *14*, 527–532. [CrossRef]
4. Dirviyam, P.S.; Palanisamy, C. Optimization of surface roughness of AISI 304 austenitic stainless steel in dry turning operation using Taguchi design method. *J. Eng. Sci. Technol.* **2010**, *5*, 293–301.
5. Xavior, M.A.; Adithan, M. Determining the influence of cutting fluids on tool wear and surface roughness during turning of AISI 304 austenitic stainless steel. *J. Mater. Process. Technol.* **2009**, *209*, 900–909. [CrossRef]
6. Kaladhar, M.; Subbaiah, K.V.; Rao, C.H.S. Optimization of surface roughness and tool flank wear in turning of AISI 304 austenitic stainless steel with CVD coated tool. *J. Eng. Sci. Technol.* **2013**, *8*, 165–176.
7. Wagh, S.S.; Kulkarni, A.P.; Sargade, V.G. Machinability Studies of Austenitic Stainless Steel (AISI 304) Using PVD Cathodic Arc Evaporation (CAE) System Deposited AlCrN/TiAlN Coated Carbide Inserts. *Procedia Eng.* **2013**, *64*, 907–914. [CrossRef]
8. Grzesik, W. Prediction of the Functional Performance of Machined Components Based on Surface Topography: State of the Art. *J. Mater. Eng. Perform.* **2016**, *25*, 4460–4468. [CrossRef]
9. Benardos, P.; Vosniakos, G.-C. Predicting surface roughness in machining: A review. *Int. J. Mach. Tools Manuf.* **2003**, *43*, 833–844. [CrossRef]
10. Liu, Z.Q. Finite difference calculations of the deformations of multi-diameter workpieces during turning. *J. Mater. Process. Technol.* **2000**, *98*, 310–316.
11. Świć, A.; Gola, A.; Wołos, D. Analysis of typical structures of dynamic systems of machining of elastic-deformable shafts with low rigidity. *Adv. Sci. Technol. Res. J.* **2018**, *12*, 1–9. [CrossRef]
12. Benardos, P.G.; Mosialos, S.; Vosniakos, G.-C. Prediction of workpiece elastic deflections under cutting forces in turning. *Robot. Comput. Manuf.* **2006**, *22*, 505–514. [CrossRef]
13. Świć, A.; Taranenko, V.; Gola, A. Analysis of the Process of Turning of Low-Rigidity Shafts. *Appl. Mech. Mater.* **2015**, *791*, 238–245. [CrossRef]
14. Świć, A.; Wołos, D.; Litak, G. Method of control of machining accuracy of low-rigidity elastic-deformable shafts. *Lat. Am. J. Solids Struct.* **2014**, *11*, 260–278. [CrossRef]

MDPI
St. Alban-Anlage 66
4052 Basel
Switzerland
Tel. +41 61 683 77 34
Fax +41 61 302 89 18
www.mdpi.com

Machines Editorial Office
E-mail: machines@mdpi.com
www.mdpi.com/journal/machines